Process Industries 1

In memory of Jeanne
For Pascal, Christian and Charles
Jean-Pierre Dal Pont

To Annick, Cyril and Jean-Louis
Marie Debacq

Series Editor
Jean-Claude Charpentier

Process Industries 1

Sustainability, Managerial and Scientific Fundamentals

Edited by

Jean-Pierre Dal Pont
Marie Debacq

ISTE

WILEY

First published 2020 in Great Britain and the United States by ISTE Ltd and John Wiley & Sons, Inc.

ISTE Ltd
27-37 St George's Road
London SW19 4EU
UK

www.iste.co.uk

John Wiley & Sons, Inc.
111 River Street
Hoboken, NJ 07030
USA

www.wiley.com

Library of Congress Control Number: 2020940663

British Library Cataloguing-in-Publication Data
A CIP record for this book is available from the British Library
ISBN 978-1-78630-442-1

The manufacturer's authorized representative according to the EU General Product Safety Regulation is Wiley-VCH GmbH, Boschstr. 12, 69469 Weinheim, Germany, e-mail: Product_Safety@wiley.com.

Contents

Foreword 1 ix
Laurent BASEILHAC

Foreword 2 xi
Vincent LAFLÈCHE

Foreword 3 xv
June C. WISPELWEY

Introduction xvii
Jean-Pierre DAL PONT and Marie DEBACQ

Chapter 1. Industries, Businesses and People 1
Jean-Pierre DAL PONT

1.1. Manufacturing, process, and service industries 2
1.1.1. Manufacturing industries 2
1.1.2. Process industries 2
1.1.3. Service industries 3
1.2. Founding fathers of the industrial enterprise 3
1.3. Anatomy of an industrial enterprise 4
1.4. Industrial strategy: the business plan 6
1.4.1. Industrial strategy of the company 7
1.4.2. Business plan 10
1.4.3. Reengineering the corporation 11
1.5. Systemic vision of the enterprise: the enterprise and flows 11
1.6. The two operating modes of the enterprise: operational and entrepreneurial 13
1.7. Governance 14

1.8. Operations abroad 15
1.9. References 15

Chapter 2. Earth, Our Habitat: Products by the Millions, the Need for Awareness 17
Jean-Pierre DAL PONT and Michel ROYER

2.1. Population explosion 18
2.2. Systemic analysis and the concept of a system 20
2.3. Earth, a complex system 22
2.3.1. Atmospheric chemistry, ozone, and climate change 22
2.3.2. Water-energy-food-climate nexus 22
2.4. Awareness, sustainable development. 26
2.4.1. Rachel Carson and sustainability. 26
2.4.2. Sustainable development 27
2.5. Products by the millions 28
2.6. Resource Earth, garbage Earth: towards a circular economy. 31
2.6.1. Circular economy. 32
2.6.2. Lifecycle assessment (LCA) and ecodesign 33
2.7. Materials science 34
2.8. Product formulation and engineering. 35
2.9. Product toxicology and ecotoxicology 36
2.10. Product packaging and ergonomics 37
2.10.1. Packaging and packing/wrapping. 37
2.10.2. Ergonomics. 37
2.11. New consumer requirements 38
2.12. Boxes. 39
2.13. References. 83

Chapter 3. Designing Chemical Products 85
Willi MEIER

3.1. Introduction. 85
3.1.1. Why is chemical product design important? 85
3.1.2. Current state of the art 87
3.2. Basic technologies. 88
3.2.1. Dimensions 88
3.2.2. Additives 89
3.2.3. Microencapsulation. 94
3.3. Products 97
3.3.1. Aspirin® 97
3.3.2. Coffee and related beverages 99
3.4. Product design 4.0 103
3.5. References 105

Chapter 4. Chemical Engineering: Introduction and Fundamentals 107
Marie DEBACQ, Alain GAUNAND and Céline HOURIEZ

4.1. Introduction: definitions, history, and challenges 107
4.1.1. Prehistory of chemical engineering 108
4.1.2. A crosscutting science serving society 109
4.1.3. Chemistry, formulation, industrial chemistry, chemical engineering, and product engineering 110
4.2. Fundamentals of chemical engineering 114
4.2.1. Thermodynamic fundamentals of chemical engineering 114
4.2.2. Kinetic fundamentals of process design 126
4.2.3. System-balances-performance approach for process design 145
4.2.4. Conclusion: ideal hydrodynamics and balances 152
4.3. Box 154
4.4. References 155

Chapter 5. Chemical Engineering: Unit Operations 157
Marie DEBACQ

5.1. Distillation 159
5.1.1. Vapor–liquid equilibria 162
5.1.2. Balances for a distillation column 164
5.1.3. McCabe–Thiele method 165
5.1.4. Technologies for continuous distillation 167
5.1.5. Conclusion on distillation 168
5.2. Fluid–solid mechanical separations 169
5.2.1. Fluid–solid interaction laws 169
5.2.2. Settling 173
5.2.3. Centrifugal and inertial separation 174
5.2.4. Filtration 177
5.2.5. Conclusion on fluid–solid mechanical separations 182
5.3. Stirring 183
5.3.1. Qualitative aspects of stirring 184
5.3.2. Quantitative aspects of stirring 186
5.3.3. Choice of impellers 188
5.3.4. Stirred tank scale-up 189
5.3.5. Conclusion on stirring 190
5.4. Heat exchangers 190
5.4.1. Heat exchanger technologies 190
5.4.2. Designing heat exchangers 192
5.4.3. Conclusion on heat exchangers 195

5.5. Reactors 195
5.5.1. Conversion rate and generalized extent of reaction 196
5.5.2. Ideal homogeneous reactors. 198
5.5.3. Non-ideal reactors 203
5.5.4. Multi-phase reactors 205
5.6. Conclusion 208
5.7. Boxes 209
5.8. Glossary. 223
5.9. References 224

List of Authors 227

Index 229

Summary of Volume 2 231

Foreword by Laurent Baseilhac

Without doubt, we have entered a new era. The upcoming change seems so significant that there is already talk of a digital revolution. In this context, many questions arise:

– What is the future of the process industry and the future of the talented women and men who are its artisans?

– Will we lose this level of excellence in the field? Or, on the contrary, will we cultivate it by reviving professions and vocations in terms of the challenges we see today?

– Is digital technology a winning bet?

– How are industries preparing for their digital transformation? What are the risks involved?

There are so many thought-provoking new challenges.

Jean-Pierre Dal Pont, Marie Debacq, and their co-authors set out to retrace some industrial trajectories that show that, at different times, industry professionals have had the capacity to overcome the challenges of their time (production, productivity, adaptation to consumption patterns, etc.). The challenges differ today, with societal and environmental markers becoming more and more significant. We understand, through the book, that the answers must probably be sought once again in terms of process engineering technologies, as well as in our ability to renew our management models implemented at the industrial level.

I am sure that readers will share this sense of urgency to tackle all the challenges of the new world. In doing so, we are cultivating a field of expression for our present and future talents and we are working on setting the conditions for a possible reindustrialization.

But let's not go too far, at the risk of betraying the minds of the authors who, at this stage, seek first to provoke the awakening of consciences; they then agree to give us a few solutions, but above all encourage us to pave paths beyond the narrow boundaries that we often draw out of habit in our professions; all this of course without denying the fundamentals that constitute its foundation.

Dear readers, industrialists, academics, students, and future actors or architects of our process industries, I invite you without further delay to dive into this complete, well-documented work, embellished with top quality industrial testimonies where the authors' inextinguishable passion for their industry is evident.

This is a way to no longer doubt the meaning of our profession, a tremendous burst of energy that writes the industrial history of our next decade.

Laurent BASEILHAC
Processes Director, Arkema
Digital Manufacturing Manager

Foreword by Vincent Laflèche

When Jean-Pierre Dal Pont asked me to write a foreword for his book, I accepted it with enthusiasm.

Not just out of friendship. We have known each other and have had the opportunity to collaborate for over 15 years. We share the same belief in the importance of strengthening links and exchanges between industrialists, researchers and teacher-researchers, between private research, academic research, and higher education.

As Deputy Director, then Chief Executive Officer of INERIS (*Institut national de l'environnement industriel et des risques*), then Chairman of BRGM (*Bureau de recherches géologiques et minières*), and, since 2016, Director of the *École des mines de Paris*, I have placed the development of research in partnership with companies at the heart of my strategy, while ensuring that our upstream research activity, funded by public subsidies, constantly feeds the scientific excellence of our teams.

The *École des mines de Paris* adopted its strategic plan in 2017. A member of the new PSL university (*Université Paris sciences et lettres*, which from the outset has been among the top 50 of the best universities in the world and the top French university), the school puts scientists and engineers in an environment as close as possible to a research activity. Over 75% of the school's activity is dedicated to research and only around 25% to teaching. Its ambition is to prepare general engineers capable of making a significant contribution to meet the major challenges of the 21st Century. Ecological transformation, with a particular focus on energy transformation, is clearly a strategic focus. The second area of focus is the digital transformation of companies, with a resolute positioning of project ownership for the school. Since the start of the new school year in September 2019, all our engineering students have common core subjects from the first year, which prepares them for Big Data processing, so as not to use words that sometimes go out of fashion quickly, such as deep learning and artificial intelligence (AI). Ensuring the scientific excellence, in particular mathematical excellence, of our engineering students clearly remains a lasting strategic marker.

The "resolute project ownership" approach means that these tools will subsequently be used in engineering projects, in contact with and paying close attention to industrial partners. The school has a long history of Big Data in geoscience to optimize drilling, whether oil or geothermal. The school was also recognized for the contribution of AI in the detection and treatment of cancers as part of its work with the *Institut Curie*, which is also a member of PSL. The challenge for our graduates is not only to know how to use these new tools, but also to know how to ask and design the industrial question in the wider framework opened up by these new tools, which requires a sound understanding of technological and industrial reality.

We train non-specialized engineers. For more than two centuries, the *mines* engineer has indeed had to integrate scientific, economic, and human dimensions, as well as management, security, openness, and solidarity that the beginnings of a professional career "basically" and inevitably inculcate. For this purpose, the school's training combines the so-called "hard" or engineering sciences, natural sciences, and humanities and social sciences (economics, management, sociology, etc.).

The work of Jean-Pierre Dal Pont and Marie Debacq could not better fit the strategy of the school, and vice versa. The words "theory and practice" have been inscribed on the pediment of our establishment for almost two centuries. Similarly, this work is punctuated and illustrated by concrete cases. This choice can only delight the Director of the *École des mines* that I am. These concrete cases relate to hot topics, often widely publicized (Smart Citites, plastic recycling, etc.). It is not for the sole purpose of following the news: it is recognition of the fact that we are going through a period where we have ever faster cycles of innovation – driven in particular by the digital revolution – and that we need innovation and new technologies to meet the challenges of sustainable development, but also that these innovations are not necessarily accepted by the general public in our society, where we are seeing trust in the engineer and the expert decrease. The engineer and the scientist must integrate this dimension into their approach. The tools and methods developed in this book perfectly integrate these challenges.

This book therefore not only makes an exciting connection between the challenges of business, research, and higher education, it also opens the reader up more broadly to major societal challenges. It does so at a time when, while ecological and energy transitions are underway, the digital revolution is bringing about profound changes to the conduct of companies and industrial management.

This revolution also has consequences on society and brings its share of fears and fantasies, like those engendered by robotics. This book comes at the right time to help students, teachers, researchers, and professionals in their choices and their reflections concerning a rapidly changing world where science and technology are increasingly essential players in sustainable development.

I highly recommend it!

Vincent LAFLÈCHE
Director
École des mines de Paris

Foreword by June C. Wispelwey

Fifteen years ago, when starting the Society for Biological Engineering of the American Institute of Chemical Engineers (AIChE®), I had the good fortune to meet Jean-Pierre Dal Pont. We talked about the future of the chemical engineering profession and about the influence of advancements in biological engineering, particularly with bio-energy and bio-pharma. We discussed the opportunities for creating new life-saving therapeutic proteins and chemicals produced economically from renewable feedstocks. This was the first of many inspiring conversations we would have regarding the future of chemical engineering. It is not a surprise to me that he wrote this passionate book about the process industries and a vision of its future.

Now is a time of transformation. AIChE has a ground floor view of one aspect that is gaining ground – process intensification and modularization. The effort, led by the AIChE's RAPID Manufacturing Institute, is dedicated to improving energy efficiency and lowering investment requirements, and removing barriers that have limited deployment of this technology. For example, process intensification can combine steps and lead to lower costs in industries such as oil and gas, pulp and paper, and chemical production. Modularization enables one to add capacity in small increments which are more suited to a manufacturer's need. The Institute de-risks new technologies in these capital-intensive industries and reduces the ecological footprint.

Another new transformative technology is digitization, or Industry 4.0. There are many aspects of digitization, including the internet of things, smart manufacturing, 3D printing, enhanced or virtual reality, artificial intelligence, big data, robotics, drones and more. These individual technologies are made possible by the new speed of computation, though they are threatened by cybersecurity. Chemical and other engineers, technicians, operators and all those who work in the process industries will need to understand and work with these technologies as they mature.

These volumes arrive at the right time for the new generation, who will enable these technologies and develop new ones to strengthen the process industries and make the world a better place.

June C. WISPELWEY
Executive Director and CEO
AIChE

Introduction

This book, a result of knowledge exchange between the academic and industrial worlds, aims to introduce **process industries** to students, teachers, researchers, professionals, decision-makers, and, in general, the general public, at a time when they are affected by the **digital revolution** that accompanies the ongoing energy and environmental transitions.

These industries aim to transform and/or separate matter by chemical, physical or biological means. They cover huge and often complex fields such as chemistry, petroleum, pharmaceuticals, cosmetics, metallurgy, food industry, biotechnology, environmental and energy industries, among others. Their economic and societal importance is considerable.

These **companies** create **value** through their products from industrial facilities (workshops, factories) that implement specific **technologies** and **processes**. The science enabling this implementation is called "chemical engineering" (*génie des procédés* in French).

The French name is to be credited to the late Professor Jacques Villermaux of the *École nationale supérieure des industries chimiques* (ENSIC, the French National School of Chemical Industries) in Nancy, who noted that all the knowledge and techniques of chemical engineering could be perfectly applied, beyond the chemical and petroleum industries, to all process industries.

This book is an invitation to discover the operational modes and technical and industrial management of these industries. It attempts to succinctly answer the following questions:

Introduction written by Jean-Pierre DAL PONT and Marie DEBACQ.

– What is a company?

– What are its foundations and how is it organized?

– How does it respond to what is today known as CSR (corporate social responsibility)?

– How does it cooperate with its stakeholders (clients, stockholders, employees, administration, etc.) when the concept of capitalism with a human face is born which, in addition to remunerating its shareholders, wants to display its contribution to the common good?

– How does it design its commercial products based on the results of its research?

– How does it build and manage its plants and factories to manufacture and distribute its products, after having assessed their impact on the environment through an eco-design analysis based on LCA (Life cycle Assessment)?

– What are the scientific bases and the "technological elements" that the chemical engineer, at the heart of the process, will use to design and operate the **manufacturing facility**?

To ensure their sustainability, process companies must adapt to their socio-economic environment, and, more particularly, to the society they shape through their innovations and products. In particular, they can help respond to the major challenges of today's world, such as that of population growth: if we believe the forecasts, there will be two billion more people to feed by 2050. Growing urbanization will also create quickly insurmountable problems if they are not managed now: a city like Chongqing, on the banks of the Yangtze, has a population that represents half of the population of France. The concepts of Smart Cities and Smart Buildings are therefore essential. As for climate change, this is perhaps the biggest challenge on the planet: the water stress associated with it will affect at least 17 countries, including India. Water is life!

Added to this is the fact that the increasingly enlightened consumer wants to know what they have on their plate, to be informed about the origin of the products they use. Traceability, authentication, naturalness, fair trade, etc. are concepts that manufacturers can no longer ignore. For example, the world is worried about the future of plastics: The Great Pacific Garbage Patch and the North Atlantic Garbage Patch[1], which are several times the size of France, are dumbfounding.

1 Continents of plastic floating on the oceans, sheltering an aquatic fauna that feeds on it and enters the food chain.

This book is particularly interested in the industrial facility at the center of the company. The future of it will depend heavily on its design and its technical and human implementation. Manufacturing operations are no longer considered dirty jobs; it is a given that wealth is built in the workshop (or on the shop floor). Thus, Toyotism, also called "lean manufacturing", is there to prove it: this production system has enabled Toyota to create an empire in the automotive industry and surpass the Americans in their own country.

In recent years, the **digital revolution** has brought about a radical change (disruption) at the societal level and at the level of companies, both at the managerial and productive levels. It was made possible by the increased power of computers (Moore's law), by the multiplication of sensors, their miniaturization, their low cost, and the development of **algorithms**. The notion of artificial intelligence (AI), which brings together a set of **computer applications** and **algorithms** based on the processing and exploitation of Big Data, testifies to this industrial revolution in progress. AI modifies our lives, our professions, our way of moving, very often, of taking care of ourselves, without our being aware of it. This term pervades books, articles, speeches and private and government research programs. Smartphones and tablets, which are only about 10 years old, are one of the essential pieces of media of this revolution. Who could do without it today?

In addition to AI, the digital revolution has brought with it a number of **digital tools** that underpin the concept of the **factory of the future**, born in Germany under the name "factory 4.0". The factory of the future combines the virtual world with the real world. These tools include the IoT (Internet of Things) – everything is connected and everything is connectable – virtual reality, augmented reality, digital twins, additive manufacturing (3D printers), etc. The world of work is deeply affected by robotics and cobotics. We must expect an industry to emerge where repetitive, tiring, messy and even dangerous tasks will be eliminated. The operator will be more of a supervisor than a performer.

Added to this is the fact that the concept of sustainable development, the basis of CSR, is now mature, including the need for metrics. Industry is moving towards a circular, low-carbon and, no doubt, decentralized economy. Bio-industries are not immune to this development with the development of synthetic biology, a remarkable future technological tool, but subject to controversy from the ethical standpoint.

In this shifting context, it is therefore difficult to grasp what the evolution of employment will be; dignified roles are created (Data Officer), while subordinate tasks are on the way out.

Are we moving towards a **civilization of algorithms**? Their opacity raises fears of the advent of a "Black Box Society" where individual freedom is in danger. Everything is known, everything can be known! Our societies – already based on science, technology and knowledge – will become increasingly connected and undoubtedly more complex and more vulnerable.

GAFA (Google, Apple, Facebook, and Amazon), the most powerful digital Internet companies in the world, are already frightening with their capital power, supranationality, and speed of deployment. In this global technological race where everything is accelerating, China has now entered the fray and faces the United States.

These are the reflections that this work invites us to. This book hopes to be interactive and accessible for everyone; it refers to illustrative videos and presents concrete examples, offered by leading figures in the form of boxes. These are listed at the end of each volume.

Videos

The following link to a website makes it easy to access the resources that illustrate this work, in particular, the videos:

https://frama.link/livreIndustriesProcedes

The links and videos are classified by volume and by chapter (via the menu on the left), in the order of appearance in the book.

Volume 1: Sustainability, Managerial and Scientific Fundamentals

Chapter 1: Industries, Businesses and People (Jean-Pierre Dal Pont): this first chapter is devoted to the industry and the businesses that depend on it. It focuses on process industries, while highlighting what differentiates them from the manufacturing and service industries. The themes concerning their constitution, strategy, functioning and governance are discussed.

Chapter 2: Earth, Our Habitat: Products by the Millions, the Need for Awareness (Jean-Pierre Dal Pont and Michel Royer): dedicated to the relationship between products and the environment, this chapter initiates a reflection on our way of life. Earth, our habitat, is a finite space whose complex cycles depend on anthropic activities: we can cite, for example, atmospheric chemistry and the problem of ozone. The vital systems of water, food, energy and climate are referred to as a "nexus", because they are interdependent. Products, whose quantity is increasing with the population explosion, must be ecodesigned using LCA (Lifecycle Assessment), toxicology, ecotoxicology and traceability studies, and turn to biobased raw materials. The circular economy must prevail over a linear economy, which consists of extracting, producing, consuming and throwing away.

Chapter 3: Designing Chemical Products (Willi Meier): Chapter 3 is dedicated to product design and formulation. A product must be designed to meet the needs of customers. In now saturated markets, companies are turning to often complex functionalized products. Post-its are a vivid example: at first, it was just a glue that stuck badly! Who could do without them today? Increasingly based on bio-sourced raw materials and biotechnologies, products use additives: ingredients such as starch and gelatin. This is the case for drugs that can also be encapsulated with alginates to reach the right target at the right time. The story of Aspirin®, first synthesized by Bayer in 1897, is remarkable. Its survival is due, in part, to sophisticated formulations. Another example of the development of drinkable formulations is coffee. The formulation of environment-friendly "smart" products in the field of textiles or fertilizers, for example, is a science with a bright future.

Chapter 4: Chemical Engineering: Introduction and Fundamentals (Marie Debacq, Alain Gaunand and Céline Houriez): chemical engineering, although omnipresent, is almost unknown to the general public. The beginning of this chapter therefore endeavors to give some definitions and historical benchmarks about this young applied science. The fundamentals of chemical engineering are then presented: starting with thermodynamics, then transfers, and finally chemical kinetics and catalysis. The last part of the chapter presents the "system-balances-performance" approach for process design using two simple examples. A box presents the very first level of calculation on processes, namely material balances.

Chapter 5: Chemical Engineering: Unit Operations (Marie Debacq): the concept of a unit operation has made it possible to bring together, in large categories, the innumerable equipment used by the process industries. There are numerous unit operations and there is no a question of giving an exhaustive presentation here. This chapter therefore covers some of them, chosen because they are particularly symbolic or representative of one type of operation or another. Thus, the following are presented: distillation, the most important separation operation and also certainly the most scientifically mature; some fluid/solid mechanical separation operations, very widespread industrially but still relatively empirical today; agitation, as a symbol of the importance of hydrodynamics (that is to say, the study of fluid movements) in chemical engineering; heat exchangers, the main representatives of transfer operations (heat exchangers dealing with the process of heat transfer); and, finally, reactors, which are at the heart of processes and responsible for the transformation of matter on the scale of the molecules themselves.

Volume 2: Industrial Management and the Digital Revolution

Chapter 1: Bio-industry in the Age of the Transition to Digital Technology: Significance and Recent Advances (Philippe Jacques): the digital revolution is profoundly changing the profession of engineers involved in bio-industries. This chapter describes the main stages of development of a product of microbial origin and how approaches related to bioinformatics, synthetic biology, systems biology and microfluidics will make it possible to amplify the development of this growing economic sector.

Chapter 2: Hydrogen Production by Steam Reforming (Marie Basin, Diana Tudorache, Matthieu Flin, Raphaël Faure and Philippe Arpentinier): this chapter presents the most widely used hydrogen production process in the world: steam reforming of natural gas. All the technological elements of this process are described, as are the problems of industrial operation of these units. Current and future developments, including those aimed at minimizing carbon dioxide emissions, are also discussed.

Chapter 3: Industrialization: From Research to Final Product (Jean-Pierre Dal Pont): the process includes all the technologies that plants and factories use to manufacture a product or a set of products. Very generally, this is a reaction followed by purification: a drug or a product to protect plants, often complex molecules, are the result of several reactions and several separations or purifications called "unit operations", described elsewhere.

The purpose of this chapter is to describe the industrialization process, which, starting from research, will define the production tool. At the end of the chapter, two boxes describe the increasingly sought-after modular construction and the constraints and advantages of a multi-workshop platform.

Chapter 4: Operations (Jean-Pierre Dal Pont): operations, or manufacturing, designate the implementation of industrial facilities (plants or factories). They are an essential function of the process industries, the source of their products and related services, and, therefore, of their profit.

This chapter studies production, its flows (financial, information, materials), and the increasingly sophisticated IT tools that make it possible to manage them such as ERP (Enterprise Resource Planning). It also discusses the bases for calculating the cost price of products and margins. Finally, special thought is given to change management: to last is also to change.

Chapter 5: The Enterprise and The Plant of the Future at the Age of the Transition to Digital Technology (Jean-Pierre Dal Pont): Chapter 5 recalls the industrial revolutions that have followed one another since the invention of the steam engine, a source of energy at the beginning of the 18th Century, to the present day. It analyzes their impact on society and on the capital-intensive business as we know it today. Emphasis is placed on information technology, which took off after the Second World War. The emergence of the Internet around 1990, that of the smartphone around 2000, and the beginnings of artificial intelligence initiated the digital revolution, whose unprecedented impact we are already seeing on society and industry. Many boxes give examples of the use of AI in fields as varied as autonomous cars, underwater exploration, robotics and industrial management.

Chapter 6: And Tomorrow... (Jean-Pierre Dal Pont): this last chapter is a reflection on the digital revolution as it is perceived today and, more particularly, on artificial intelligence, which is its standard-bearing media. AI is increasingly affecting the city which wants to be *smart*. The water sector is taken as an example of economic activity whose digital aspect modifies the processes, the management of the distribution networks and the trades.

While the various applications of this emerging technology can hold out hope for many advances and improvements, the use of AI raises many questions. The very functioning of the industrial business is turned upside down. Will Teslism, synonymous with, among other things, the "hybridization" of computer systems, supplant Fordism? Isn't the robot assisting the operator a threat to his job? The citizen, meanwhile, questions the intrusion of GAFA in his private life and governments about their supranationality. The "fully connected" raises fears for the fragility of administrative and industrial systems, while cybercrime is a ubiquitous threat. The fundamental question is whether human beings are at the heart of the system and… for how long.

In the current period of upheaval where "the only certainty is uncertainty", perhaps we must take one of the thoughts of the great manager of the 20th Century, Peter Drucker: "The best way to predict the future is to create it." One of the ambitions of this book is to help the readership in this research, or at least, to try to whet their curiosity.

1

Industries, Businesses and People

A company, in the broadest sense, is the world of work, of employment. It transforms societies and landscapes through technological revolutions. Its modus operandi is complex and influenced by the political system, the cultures of the countries that host it, and the ways in which it is financed.

In what follows, we give a very brief description of the "upscale" enterprise of a developed country like France, using previous publications. Talking about the impact of the digital revolution on business requires a succinct description of what the latter is: this is what we have tried to do.

Behind the term "industry" is "industrious", the quality of people who have the idea of producing material goods or products from raw materials that are processed for sale. At its core, it is deploying capital to seek gain, profit, and activity.

One to two million years ago, our distant ancestors had already discovered materials, including flint, a very hard rock, to make tools – extensions of the hand – and weapons (arrowheads, axes). Our ancestors, like today, sought to survive: we must eat and defend ourselves.

Let's skip forward a century or two! During the Industrial Revolution, discussed in Chapter 5 of the second volume, handicraft was replaced by industry, characterized by raising capital, the use of machines that made use of the new energy, that is, steam, and men mostly gathered in factories. A century earlier, Colbert (1619–1683) had created a number of factories during the reign of the Sun King (King Louis XIV of France); it was a question of making manufactured products which were equivalent to products made by hand, although sometimes made using machines (looms).

Chapter written by Jean-Pierre DAL PONT.

Mechanization, mining, and capitalism, the history of which Joyce Appleby described in a masterful book (Appleby 2010), emerged during the Industrial Revolution. She defines **capitalism** as:

> a system based on individual investments in the production of marketable goods.

At the end of the 19th Century, the following were developed: railways (the East Coast of the United States joined the West Coast in 1869, in Promontory, Utah), maritime transport, inventions that succeeded each other at high speed in all areas and led to the birth to the modern enterprise. The main stages of this evolution can be found in (Dal Pont 2012); to simplify, we will say that the enterprise in which we are interested and which we will describe as modern descends from the manufacturing company of the end of the 19th Century.

The company is a source of wealth, synonymous with employment. Wealth is created in the workshop; this is the thesis defended by Adam Smith (1723–1790), British and founder of modern economics in his 1776 work, *The Wealth of Nations*.

In what follows, we will distinguish three main classes of companies.

1.1. Manufacturing, process, and service industries

1.1.1. *Manufacturing industries*

We define these as activities whose production is said to be discrete, because they offer identified objects. It is the field of the automotive and aeronautics industries, and household appliances.

They often call on the chemical industries: an automobile, for example, is riddled with chemicals (windshield, dashboard, seats, pipes, sheath of electric cables, etc.).

1.1.2. *Process industries*

These transform raw materials and/or energy by chemical, biochemical, or physical means. We can cite chemistry, pharmacy, cosmetics, metallurgy, etc. Their economic importance is considerable. Chemistry is the "mother of all sciences", said Thomas Edison; we can say that it is also the mother of all industries.

Process industries can have discrete parts: the active ingredient of a drug can be manufactured in the form of a (discrete) tablet in boxes, boxes in crates, crates in pallets. Bag-filling machines, wrapping machines, and packaging machines are often the last link in the process industries. It's mechanics, not chemistry!

1.1.3. *Service industries*

These normally have fewer industrial assets. They serve as support for producers: maintenance, engineering, design office, transport, catering, mail, etc. SNCF (France's national railway company) is a service company that uses huge industrial tools.

We can see that the line is not exactly straight. The important thing is to identify the professions involved, the skills, and the customers. It is not so obvious when we delve into the question.

1.2. Founding fathers of the industrial enterprise

The industrial enterprise as we see it today took shape at the end of the 19th Century. Among all the inventors and entrepreneurs who shaped it, we have chosen four giants from the 19th and 20th Centuries. These men contributed to an unprecedented industrial boom and profoundly changed the society of the time. We could no doubt have chosen others, but, as we know, "to choose is to renounce".

Detailed explanations on the contribution of these four great figures to the industry can be found in (Dal Pont 2012). To summarize:

– **Frederick W. Taylor**, father of *Taylorism*, analyzed the work of the workers at the workshop and evaluated each gesture in time and motion to increase productivity. A specific job was shared – divided – between several individuals. Humans are enslaved to the machine: who does not remember Charlie Chaplin's *Modern Times*? Taylor was *behind* workers with a stopwatch; he sometimes owed his salvation to running away;

– **Henri Fayol**, father of *Fayolism*, invented the functions of management, a model still widely in use today (see section 1.3);

– **Henry Ford**, father of *Fordism*, was a strong supporter of *Taylorism*. A farmer's son, he invented the assembly line and standardization, which made the automobile a cheap product, as well as the industrial integration. His Model T, "a motor car for the great multitude" in 1908, was full of inventions (transmission, suspension, etc.). It was the car of the century!

– **Thomas Edison**, the legendary inventor with 1,093 patents, created laboratories *to do business*: as soon as the luminescent lamp left Menlo Park in 1882, it illuminated Pearl Street in New York, thanks to the direct current generated by its first power plant. We can say that *Edisonism* is the way to create business by doing research, R&D, or research and innovation, as we would say today.

NOTE.– Why not Gustave Eiffel? Inventor and businessman, Gustave Eiffel is the archetype of the 19th Century engineer. He left works that are still admired by the masses. Should we not talk about the Eiffel Tower? However, he did not affect and transform the society like the "nominees" mentioned above did.

Figure 1.1. *Portraits of the fathers of the modern enterprise (from left to right) Frederick Winslow Taylor (1856–1915), Henri Fayol (1841–1925), Henry Ford (1863–1947), Thomas Edison (1847–1931)*

1.3. Anatomy of an industrial enterprise

What follows is not a lecture. Our goal is only to provide a few benchmarks to students and professionals to help them in their reflections on the world of business and perhaps guide them in their professional choices.

A business rests on four pillars:

– economic;

– financial;

– human;

– legal.

It is an organization that aims to offer goods and/or services. One or more people have raised capital and risked their money in the hope of earning it. **A business is a human adventure. Managing a business means managing risks.**

To exist, the enterprise must have a **legal status**, which therefore makes its president responsible before the law. There are many legal forms: public limited company, SARL (*société à responsabilité limitée*, a private limited liability company), non-profit association, as per the 1901 law, etc. Its foreign subsidiaries must comply with local laws.

In France, size differentiates **VSEs** (very small enterprises or businesses: less than 10 employees), **SMEs** (small or medium enterprises or businesses: less than 250 employees and a turnover of less than 50 million euros), and **medium-sized enterprise** (between 250 and 4,999 employees and a turnover not exceeding 1.5 billion euros). Beyond that, it is the domain of very large companies and multinationals.

A company's headquarters has certain functions: Fayolism is still relevant! These functions are often shown by a company's organizational structure. This generally includes some or all of the following entities:

– a headquarters that brings together the president, managing director, executive management, communication, human resources department (HRD), financial department (taxes, treasury), planning department (strategy), legal agreements, research department, industrial management, engineering management, sales management, marketing management, project director(s), safety and security management, branch managers, SBU[1] manager(s), purchasing department, administration (accounting, payroll, IT), country delegate(s)[2];

– one or more production facilities (plants, workshops);

– distribution (logistics, warehouses, etc.);

– research means (laboratories, test means);

– technical services, after-sales service, application laboratories;

– commercial services (sales reps, etc.);

– subsidiaries (national, foreign);

– JVs (joint ventures).

A small to medium-sized company that includes only one site (headquarters + plant) will have all these functions distributed between a few individuals donning multiple hats:

1 Strategic Business Unit. It is an activity; glues, paints, organic products, etc. It is often a company within a company, because it has its own resources.

2 In some countries, large multinationals have a representative who facilitates contacts, informs the parent company, etc.

– so-called corporate functions are called *on staff* (legal agreements, safety, etc.) when they support so-called *on line* functions such as manufacturing. We can note the importance that communication has taken over the years;

– in general, an executive committee (board) of a few people, including the president and the general management, takes the essential decisions, the directions that will be implemented by the lower levels;

– the HRD is often the first contact for anyone looking to join a company;

– as we will see below, the digital revolution is creating new roles; Data Officer, Data Scientist. These roles may report to the president, to a board member or to the industrial management. It all depends on the importance that the company gives this new function at a given time.

Start-ups (short for start-up companies) work on something new; an innovative object, a new drug that leaves hope for big profits when research comes to an end. They look for funding to survive. Another term has appeared recently, mainly in the digital field: that of a "unicorn", a company whose market capitalization reaches one billion dollars. The famous Silicon Valley saw the prodigious birth of these mega-companies, launched in a garage or a student room.

1.4. Industrial strategy: the business plan

Any business must have a long-term vision, to plan for the future. It must examine and review its strategy, which can simply be defined as an objective to be achieved, and determine the related means. In what follows, we will limit ourselves to the industrial strategy, which amounts to examining the **adequacy** between the products that the company has at a given time and the needs of the customers, which can be very variable from country to country.

A product is born, lives, and dies, as shown in Figure 1.2a. The turnover of a company, and therefore its profit, is composed, at a given moment, of products that are at different stages of their existence, as shown in Figure 1.2b. New products in the process of being launched sometimes cost money instead of bringing in money, because the related commercial costs (product managers, advertising, canvassing) are not covered by their contribution margin, as these products are manufactured in too small a volume. For the same reasons, end-of-life products are no longer competitive: they have found replacements, either in the company that created them, or in the competition; their selling prices tend to fall.

In general, a company has flagship products from which it hopes to "benefit" for a long time. Pharmaceutical companies dream of blockbusters whose turnover exceeds one billion dollars. The vitality of a company, the **efficiency of its research**, can be measured by the percentage of profit brought by products that are a few years old, and are therefore young products. Often, financial analysts try to assess the products that the company has in the "pipeline".

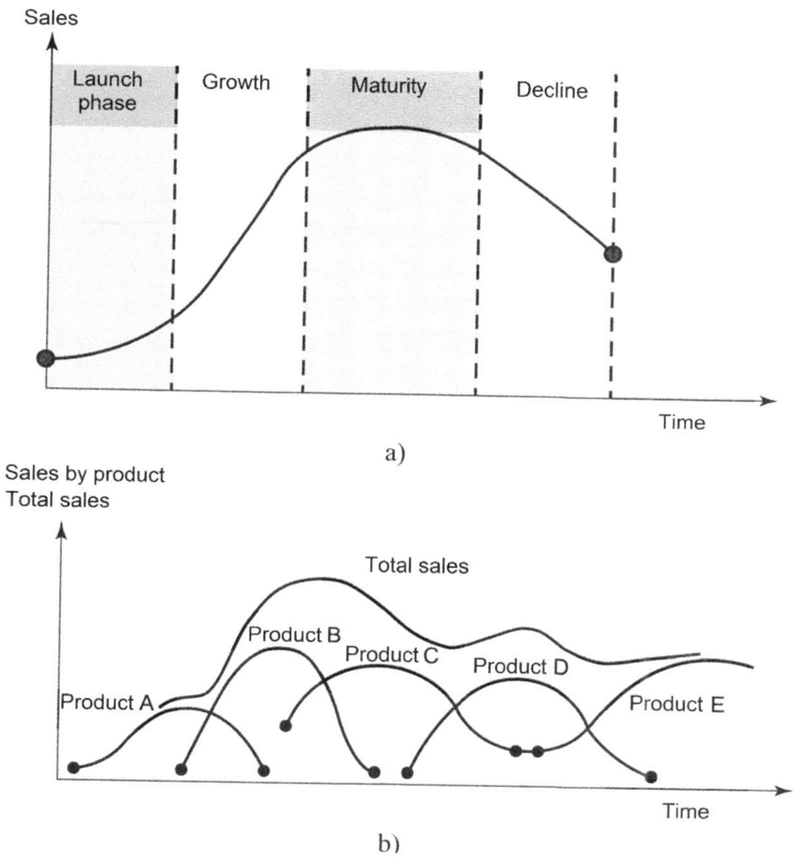

Figure 1.2. *a) Product life cycle; b) company turnover versus products*

1.4.1. *Industrial strategy of the company*

The analysis of the company's industrial strategy is based on two elements:

– business analysis of the product/market couple as described in Chapter 2;

– evaluation of the industrial facility.

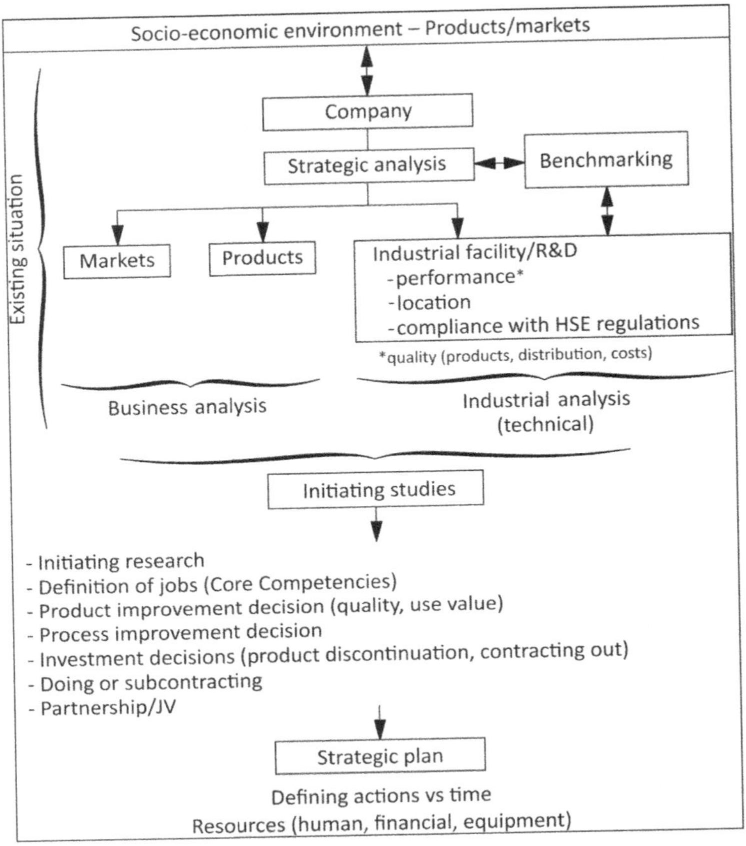

Figure 1.3. *Principle of strategic analysis of a company (technical aspects). Establishing the strategic plan*

1.4.1.1. *Business analysis*

Business is king! France's *Trente Glorieuses* (The Glorious Thirty, 1945–1975), when simply producing was enough, seems a long time ago; products were sure to sell, since there was a scarcity at the time. It was a period of a production economy. Today, consumers find everything they want. Competition has full hold: this is the market economy.

The commercial aspects of a product can be evaluated for each sector in all the countries where it is sold. A biopolymer like xanthan gum fits into several businesses: food, cosmetics, paints, or even drilling fluid, where it is used for its texture, thickening and gelling properties.

Business issues are numerous:

– market aspects:

- is the product in the markets in sufficient quantities (penetration rate), does it cover the demand, are there other markets?

- should it be taken to other countries?

– technical aspects:

- does the product have the expected functionalities?

- is it pure enough or too pure?

- is it well formulated, well-conditioned, well packaged? What about its logistics?

- does its cost price bring acceptable margins or can it not be sold because it is too expensive when it comes out of the factory?

1.4.1.2. *Analysis of the industrial facility*

The industry and manufacturing are at the service of the business; that is their *raison d'être*. In the chapters that follow, we will lay the groundwork for evaluating the processes and the plants that implement them. At this point, we will limit ourselves to substantive questions:

– is the plant a good one? In other words, how does it compare to the competition in terms of efficiency, profitability, and management (safety, stability, branding, and compliance with regulations)? Do we have any idea of its **durability**? All of these issues fall within what is called **benchmarking**;

– is it in the country where we want it to be? Does it have the right capacity?

Obviously, the cost price of a product is the primary element of interest to the salesperson in charge of its sale. To this essential notion must be added the consistency of quality and reliability of production.

A **strategic plan** consisting of a number of potential actions will result from the confrontation between business and industrial analysis (see the bottom of Figure 1.3). These actions are:

– executive management: investments (internal growth), divestments, JV (joint ventures), divestment of activities, search for new activities, doing or making others do (subcontracting), partnerships, corporate purchases (external growth), etc.;

– marketing: search for new products, modification of old ones in terms of formulation (see, in Chapter 3, the case of aspirin, a century-old product that has been rejuvenated by adding additional functions, such as effervescence), innovation (innovation is sometimes making the old new again!);

– research: new synthesis processes, new products, new applications;

– manufacturing: improved operations, construction of new plants, closure of unprofitable plants.

The definition of **core competencies** is a difficult exercise that we will deal with through KM (knowledge management).

Development abroad, which has gained considerable importance through globalization, has its own unique characteristics, which we will see later.

1.4.1.3. *Definition and implementation of a strategic plan*

All of the above thoughts and ideas need to be structured and defined in terms of human and financial resources over time. The mobilization of the company's executives and hiring consultants and engineering companies to do feasibility studies costs money. Is it necessary to remember that the resources of any organization are limited?

The following keywords can help prioritize actions, that is, set their launch date. The term "**prioritization**" is increasingly used, although it grates on the ears of many.

1.4.1.4. *Pareto analysis revisited*

In order to compare various scenarios, the following keywords can be used: expected return, chances of success, ease of implementation, risks, necessary means, time factor, prioritization. Each keyword is assigned an "importance", a "weight" ranging from 1 (low) to 5 (high). Common sense, professionalism and the use of the *collective intelligence* of the company generally make it possible to eliminate irrelevant projects. Priority will be given to a project that has a good chance of succeeding quickly with limited resources, as opposed to projects that appear fabulous but require a lot of resources and time.

Actions on which the survival of the company may depend are called **strategic**; these are, for example, very large investments and major acquisitions.

1.4.2. *Business plan*

This name generally refers to the dossier for requesting financing from a bank in the context of setting up a business. It contains commercial and financial aspects; it defines the structure and organization of the new company.

1.4.3. *Reengineering the corporation*

Reengineering the Corporation (Hammer and Champy 1993) and *Liberation Management* (Peters 1992) are the provocative titles of two American books published in the 1990s. The idea was to shake up corporate America that had rested on its post-war laurels.

The following quotes, taken from these two books, give an idea of this "crusade":

– "forget what you know about how business should work. Most of it is wrong!";

– "American companies function according to principles that are two centuries old and mainly according to the principle of the division of labor put forward by Adam Smith in 1776";

– "we live in a world governed by the three Cs: customers, competition, change";

– "reengineering is the fundamental rethinking and radical redesign of business processes to achieve dramatic improvement such as cost, quality, service and speed";

– "*reengineering ≠ restructuring ≠ downsizing ≠ reorganizing ≠ flattering an organization ≠ automating*";

– "Why do we do what we do? Why do we do it the way we do?";

– "People, jobs, managers & values are linked together. Processes, not organizations, are the object of reengineering."

When we look at what has happened to the American automotive industry in the face of Japanese competition, it is not certain that the message was heard!

The term reengineering can be considered as the awareness of a company that is asking fundamental questions about its existence. It therefore calls on high-level consultants to help it review its strategy and its way of working from top to bottom. One can imagine the state of mind of the employees who are experiencing reengineering and who wonder if the remedy will not kill the patient.

1.5. Systemic vision of the enterprise: the enterprise and flows

Figure 1.4 gives a systemic vision of an enterprise that has invested in an industrial facility. The term "CAPEX" (Capital Expenditure) is now used for capital investment. The expenses relating to the operation costs of this tool are abbreviated "OPEX" for Operating Expenditure.

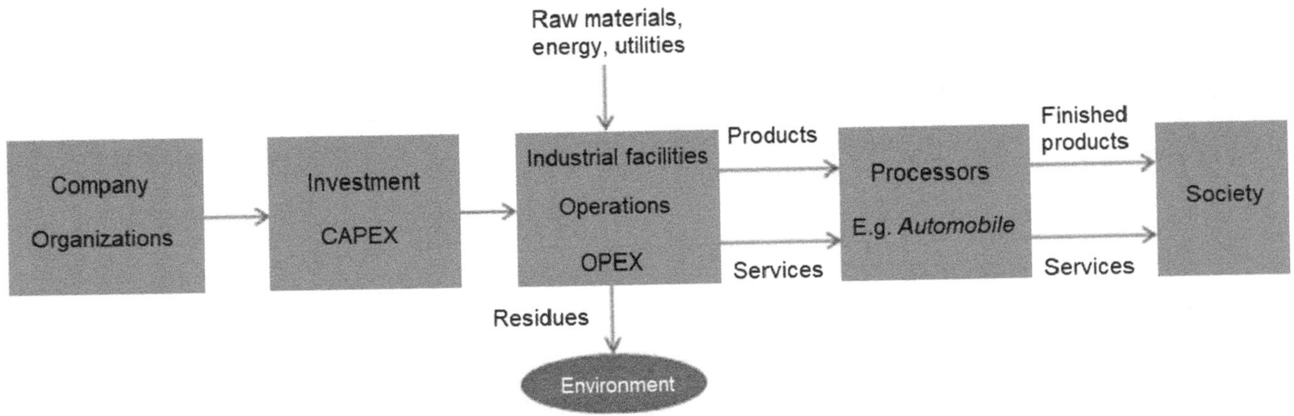

Figure 1.4. *Systemic vision of the enterprise*

Figure 1.5 complements Figure 1.4.

The enterprise is seen as its material, information, and financial flows. We have added the flow of people, in order to demonstrate the importance of employee turnover, beyond the normal erosion caused by retirements. A high turnover is often synonymous with social problems, dissatisfaction; it is a source of instability. **Managing a business means, among other things, mastering all of these flows.**

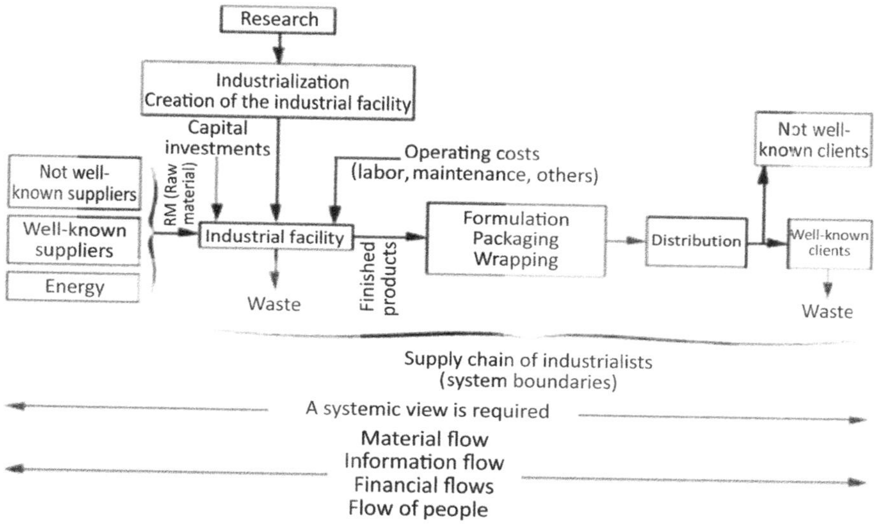

Figure 1.5. *The enterprise and its flows*

1.6. The two operating modes of the enterprise: operational and entrepreneurial

The enterprise lives at the rate of two different modes by their nature, their time constant, and the people involved.

The operational mode amounts to managing the daily, the existing, managing the workshops and plants as they are at a given time, in accordance with the values of the company and its governance (see section 1.7). This role is primarily assigned to plant directors, production managers, and to the manufacturing function in general.

The entrepreneurial mode consists of managing the **portfolio of projects** generated by the strategic plan as defined above. This is the privileged field of research, design offices, engineering, and project managers. People belonging to these entities must ensure changes that result from a strategic plan.

A major investment or reengineering can shake up a plant or system beyond reason. It is up to the executives to ensure the necessary balance. Many asset managers forget everyday life because they are attracted by the "brilliance" and novelty of investment projects.

1.7. Governance

Governance is not governing, which consists of making people do what one wants them to do and which falls within the scope of management; **governance is a set of values.** Management can be defined as running a *household*. The components of management are: planning, organizing, activating, controlling (Dal Pont 2012, p. 54). The concept of corporate governance is relatively recent and is derived from taking into account the concept of **sustainable development** in its operation.

Governance can be defined as a set of **good practices**, of procedures that will make it possible to manage a company in accordance with **ethics**, with a set of core **values**. The main governance tools relate to:

– **quality**, i.e. customer satisfaction;

– **health**, **safety**, and **environment** (HSE).

The ISO 26000 standard deals with this notion, which obliges large companies to be more transparent, in particular, in their annual report (see Figure 1.6).

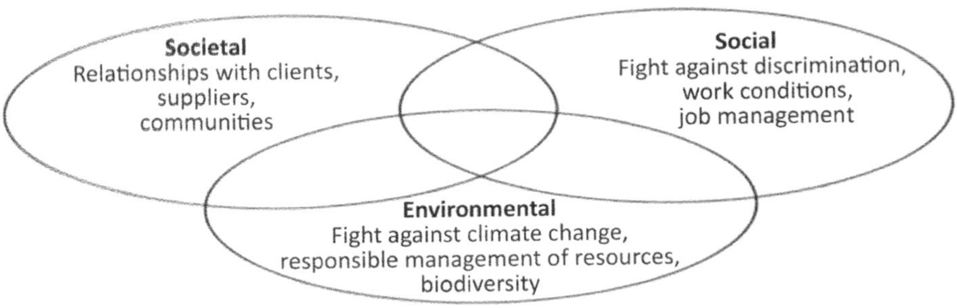

Figure 1.6. *Schema of the ISO 26000 standard*

NOTE.– Corporate social responsibility, or CSR, is described by the ISO 26000 standard. CSR is defined as the voluntary contribution of companies to the challenges of sustainable development, both in their activities and in their interactions with their stakeholders.

It concerns three areas: environmental, social, and societal. This last domain concerns the relations of the company with its partners, in particular commercial, seen from the angle of sustainable development.

1.8. Operations abroad

Trade has existed for millennia: the Phoenicians crossed the Mediterranean; the Silk Road is still something to aspire to. European and American companies have invested outside their borders for more than two centuries, creating imposing infrastructures (the Panama Canal).

Globalization supported by the Internet, containerization of transport, oil and coal flows, to name a few, and the rise of China, which is on the way to becoming the world's leading economic power in a few decades, have been game-changing. Surprisingly enough, little is said about India, soon to be the most populous country in the world.

Large companies have had to invest massively in production facilities abroad, taking advantage of their technological advance in certain fields, supporting the economic development of their local partners with whom they now compete. Recently, China has been investing abroad, not only in French vineyards, but also in technological companies: should this be seen as a fair return?

These changes, in the context of our work, give rise to reflections concerning the establishment of a company abroad, its mode of technology transfer, and the expatriation of its employees. We will return to this in Chapter 4 of the second volume.

1.9. References

Appleby, J. (2010). *The Relentless Revolution: A History of Capitalism*. W.W. Norton & Company, New York.

Dal Pont, J.-P. (2012). *Process Engineering and Industrial Management*. ISTE Ltd, London and John Wiley & Sons, New York.

Hammer, M., Champy, J. (1993). *Reengineering the Corporation: A Manifesto for Business Revolution*. Harper Business, New York.

Peters, T. (1992). *Liberation Management: Necessary Disorganization for the Nanosecond Nineties*. A.A. Knopf, New York.

2

Earth, Our Habitat: Products by the Millions, the Need for Awareness

The conquest of space resulted in an unexpected event: an astronaut of the Apollo 11 flight of July 16, 1969, which put men on our satellite for the first time, exclaimed: "We believed to have conquered the Moon, we conquered the Earth." Our habitat, the blue planet (Figure 2.1), appeared to him in all its splendor and all its fragility: habitable, covered with clouds, surrounded by an atmosphere, with continents bathing in oceans that cover three quarters of its area. What a contrast with the aridity of the moon! It appeared to these privileged observers that the Earth was a **finite space**.

Figure 2.1. *Our habitat as seen from space. For a color version of this figure, see www.iste.co.uk/dalpont/process1.zip*

Chapter written by Jean-Pierre DAL PONT and Michel ROYER.

It all started about 14 billion years ago, with the Big Bang, when space-time was created. Life on Earth appeared 3.8 billion years ago in the form of micro organisms; miraculous because life is so complex. Animals similar to humans existed 2.5 million years ago, according to Yuval Noah Harari (Harari 2014). The cognitive revolution started 70,000 years ago and the agricultural revolution 12,000 years ago. Our scientific revolution dates back 500 years.

Homo sapiens (modern human being) with a morphology identical to ours appeared between 400,000 and 500,000 years ago. They settled down ten thousand years ago: the hunter-gatherers, as free as the air, domesticated wheat, started a consumer society, says Yuval Noah Harari, not without malice. They created needs and lost their freedom!

The Neolithic (age of the polished stone) began very roughly, depending on the region, between 8,000 and 5,000 years ago. Agriculture and animal husbandry began in this period. Around 3,300 BC, the Sumerians used writing.

But let's skip some centuries ahead! Perhaps the most important phenomenon on Earth is the population explosion.

2.1. Population explosion

It took a century for the population to increase from one to two billion, from 1830 to 1930 (see Figure 2.2).

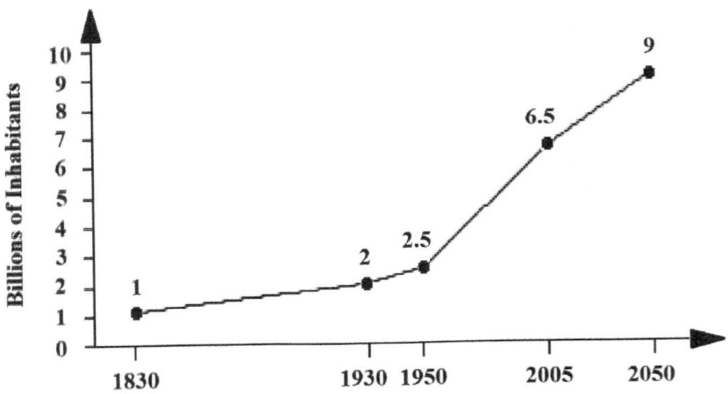

Figure 2.2. *Population explosion*

The world population will reach 9.8 billion in 2050 (+31% compared to 2017), distributed as follows[1]:

– 53.3% (5.24 billion) in Asia (+16 % compared to 2017);

– 26 % (2.57 billion) in Africa (+100 % compared to 2017);

– 12.5% (1.23 billion) in America (+22% compared to 2017);

– 7.5% (736 million) in Europe (-1.2% compared to 2017);

– 0.64% (63 million) in Oceania (+50% compared to 2017).

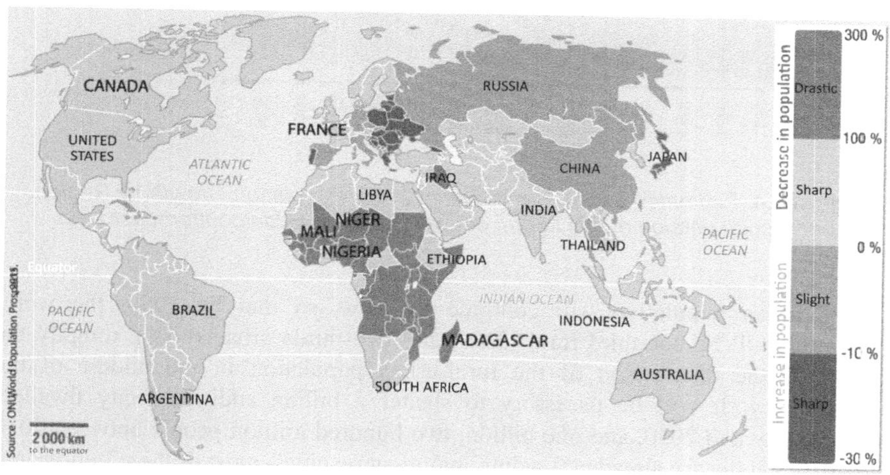

Figure 2.3. *Population change by country by 2050. For a color version of this figure, see www.iste.co.uk/dalpont/process1.zip*

Note the staggering increase in Nigeria, which goes from 197 million in 2017 to 411 million in 2050 and 794 million in 2100 (3rd position in 2100, behind India and China, and ahead of the United States, at 447 million). We note that five African countries (Nigeria, DR Congo, Tanzania, Ethiopia and Uganda) representing nearly two billion inhabitants, will, in 2100, be among the 10 most populous countries in the world[2].

In Europe, 25% of the population is already 60 years old or more. This proportion will reach 35% in 2050 and remain around this level in the second half of

1 Source: INED, September 2017.

2 Source: UN, June 2017.

the century. Populations from other regions are also expected to age significantly over the next few decades, and this trend will continue until 2100[3].

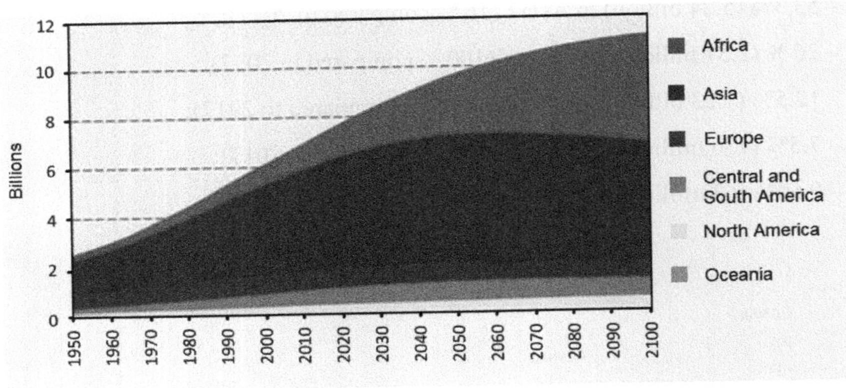

Figure 2.4. *Estimated World Population in 2100: Population Growth by Region. For a color version of this figure, see www.iste.co.uk/dalpont/process1.zip*

The urban population will continue to grow, so that by 2050, the world population will be one-third rural (34%) and two-thirds urban (66%), roughly the reverse of the distribution of the rural-urban population in the middle of the 20th Century. It will be necessary to shelter a billion additional city dwellers between now and 2030, and one billion, two hundred million people between 2030 and 2050. People are already crowding into gigantic cities, most of them unprepared for this surge. Chongqing, the largest city in the world, on the banks of the Yangtze, is home to 34 million inhabitants, nearly half the French population, and occupies an area equivalent to that of Austria: it is only a forest of buildings.

Managing a city requires a system approach, whatever its size.

2.2. Systemic analysis and the concept of a system

Von Bertalanffy (Dal Pont 2012, p. 447) is considered to be the father of systemic analysis.

A system is defined by Joël de Rosnay as a "set of elements in dynamic interaction, organized according to a goal". For example, a transport system can include a subway, buses, cars, trams, self-driving cars, cable cars, elevators, escalators, bus lanes, taxis and bicycles.

3 Source: UN, 2014.

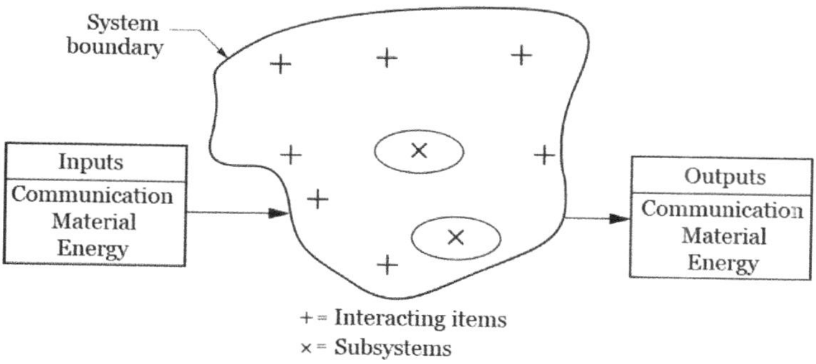

Figure 2.5. *Schematization of a system*

This simple definition hides some difficulties: it is necessary to define the limits of the system – the boundary. It is also necessary to define its elements – the items. What is really interacting? A system exchanging with the outside: information, energy, and materials. These are the inputs and outputs. As an example, Table 2.1 gives the essential elements of a city's systemic analysis: given their size, the elements are in fact sub-systems.

Inputs	City	Outputs
People Energy Food Water Raw materials Chemical products Manufactured goods Money Information Houses	Systems: hospitals, education, banking, transport, roads, energy, etc. Security: police, firefighters, etc. Leisure: parks, gardens, theaters, cinemas, tourist places, etc. Storage: cars, food, water, household products	People Residues, garbage Chemical products Pollution: air, noise, polluted water, GHG (greenhouse gases) Manufactured products Culture Various riches

Table 2.1. *Simplified systemic analysis of a city: inputs and outputs*

It is in the interest of the municipality that wants to make its city a Smart City to carry out a systemic analysis before going after this promising but complex concept (see Box 2.1).

Be that as it may, Human beings are *condemned to live*; since the beginning of time, they have known how to adapt to extreme conditions, be it the ice of the poles, the aridity of the deserts, high altitude or tropical forest. But to survive, they need water, food and energy, if only to cook their food. Given the pressure that human beings are now exerting on their habitat, they must know this at the risk of going towards disasters that some consider to be imminent.

2.3. Earth, a complex system

Our planet is the home of balances and complex cycles on which life as we know it today depends (Dal Pont 2012). To illustrate this, we will say a few words about atmospheric chemistry and the water-energy-food-climate nexus.

2.3.1. *Atmospheric chemistry, ozone, and climate change*

The chemical compounds in our atmosphere are where many free radical reactions take place; these reactions depend, among other things, on anthropogenic activities (greenhouse gas emissions), volcanic activity (emission of SO_2 and ash), and particle bombardment, particularly those emitted by the sun.

Among the molecules in our atmosphere, ozone is an important molecule. A hazardous product, it has, however, a beneficial role, because it absorbs UV rays in the upper atmosphere and thus protects the Earth from the harmful effects of this radiation. In 1995, the Nobel Prize in Chemistry was awarded to Paul Crutzen, Frank Sherwood Rowland, and Mario Molina for discovering that nitrogen oxides and chlorofluorocarbons destroy stratospheric ozone, the true protective screen of our planet (Daniellou and Naël 1995).

Joe Farman, a British geophysicist specializing in Antarctica, and two colleagues, Brian Gardiner and Jonathan Shanklin, discovered the famous "stratospheric ozone hole" in Antarctica. A global awareness lead to the Montreal Protocol of 1987, which limits the use of chlorofluorocarbons used in cold machines as thermal fluids and as propellants in aerosols. This measure had positive effects, since today, the ozone seems to have returned to a normal level in these inhospitable regions.

2.3.2. *Water-energy-food-climate nexus*

Water is life. Without water, there is no agriculture or energy, and thus no industry. This is undoubtedly the major challenge to which solutions will have to be found in the near future, at the risk of generating conflicts and population migrations with all the problems of misery and injustice that are their common lot.

Le Monde of May 21, 2019, reported the water shortage of five capital cities: Jakarta; Cairo, which draws water from the Nile that is polluted by pesticides;

Mexico City, built on a lake that is now dry; Beijing, capital of a country with little water; and London, victim of its rampant population growth. The states bordering the Nile, the Mekong, and the Jordan compete for this precious commodity; the same is true for certain water tables astride borders. The set of around 20 dams in Anatolia, called GAP, allows Turkey to control the Tigris and the Euphrates, sources of life for Mesopotamia ("the land between two rivers"), the cradle of our civilization.

Figure 2.6 schematically represents the **water cycle**.

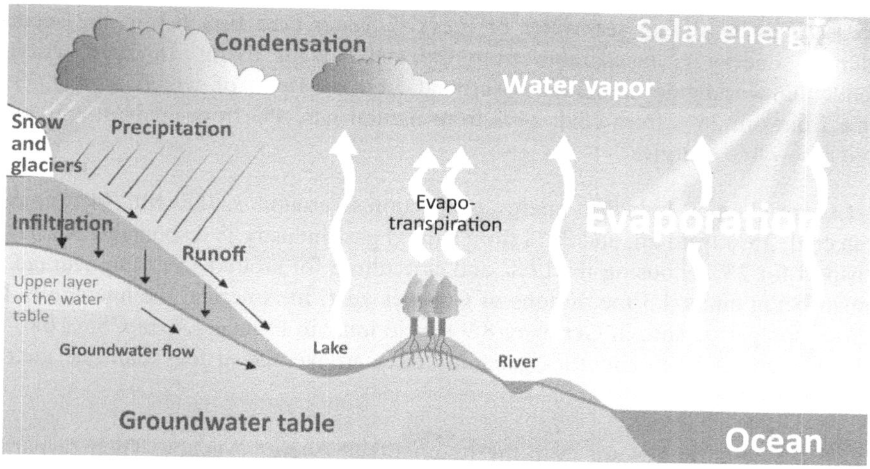

Figure 2.6. *Water cycle (source: Wikimedia Common[4]). For a color version of this figure, see www.iste.co.uk/dalpont/process1.zip*

Water is in perpetual motion under the action of the sun. Evaporated from the seas and oceans, freed of its salt, it forms the clouds moved by the winds and falls in the form of rain, snow, or hail. It feeds underground water tables, lakes, and streams.

Collected and used by humans, it becomes the vector of pollution of which the seas and oceans are the receptacle. Rainwater is generally unfit for consumption, because it has "washed" the roofs of homes and the pesticide-laden air over agricultural land. Water also "washes" the roads polluted by nanoparticles generated by cars with gasoline or diesel engines and the abrasion of tires.

Box 6.1 of the second volume gives an account of water management issues at a time of digital transition and the importance of new techniques in the preservation

4 Available at: https://commons.wikimedia.org/wiki/File:Cycle_de_l%27eau.png.

and processing of this vital raw material. Pollution, excessive use, and piracy by unscrupulous neighbors of water tables, which take thousands of years to regenerate, is perhaps one of the most disturbing aspects of water management.

The water-energy-food-climate nexus was highlighted by the World Economic Forum Water Initiative, which published the book *Water Security. The Water-Food-Energy-Climate Nexus* in 2011. This extremely complex nexus requires a localized systemic vision.

The supply, processing and distribution of water require energy, especially if it is necessary to desalinate sea water or brackish water (see Box 6.1 in the second volume). Energy is inseparable from the standard of living. In 2018, energy production worldwide amounted to around 14,000 Mtoe[5], of which around 33% came from oil, 28% from coal, 24% from natural gas, 4% from nuclear, and 11% from renewable energies[6].

CO_2 production linked to energy production is around 32,000 Mt: 44% comes from coal, 35% from oil, and 20% from natural gas. Industry is responsible for 37%, transport for 23%, housing for 17%, and agriculture for around 15%. On average, a human being emits 4.3 metric tons of CO_2 per year. In America, the average is 15 metric tons per person, in Germany 8.9 metric tons, in France 4.4, in China 6.6, in India 1.6, and African countries generally have an average of less than one metric ton per person.

All previous figures are from the International Energy Agency. Their disparity reflects both the disparity in living standards and the diversity of energy sources. We can note the impact of nuclear power, which produces almost 80% of French electricity, as well as the impact of coal on the results of China and Germany, two countries that are heavy consumers.

Climate change is currently the problem that is most debated by the world and divides it. Many countries are worried by melting glaciers, the retreat of the Arctic sea ice, violent hurricanes, cyclones, torrential rains, droughts, which are becoming chronic in many countries and record temperature. Rising waters are a specific concern for countries just above the sea level, such as Bangladesh and the Maldives.

The role of the IPCC (Intergovernmental Panel on Climate Change), founded in 1988, is to synthesize and evaluate studies relating to climate change. It attributes global warming to greenhouse gases (GHG).

5 Million metric tons of oil equivalent.

6 Renewable energies: hydroelectric (dams), wind (land and sea), solar (photovoltaic, thermal), biomass, marine (tidal, wave, thermal, etc.).

GHGs (CO_2, methane, nitrogen oxides, water vapor and other molecules) absorb infrared radiation emitted by the earth and contribute to the greenhouse effect responsible for global warming. Carbon dioxide (CO_2) resulting, in particular, from the combustion of fossil products (coal, oil, natural gas) is the GHG responsible for three quarters of the phenomenon. Coal is a big polluter. Not all the CO_2 released by human activities accumulates in the atmosphere. Part of it – around half – is absorbed by the oceans and vegetation (source: article by Damien Altendorf in *Science*, April 10, 2019). By absorbing part of the CO_2 from the combustion of fossil fuels, oceans help regulate the climate on a global scale.

In fact, the energy balances on the planet's surface are extremely complex and require in-depth knowledge that this short chapter cannot deal with. We can cite, for example, the role of **deforestation** and the **combustion of wood**, which reduces the role of forests in storing carbon, and, more particularly, the Amazonian rain forest, the real lungs of the planet. **Agriculture**, which uses considerable fertilizers and machinery, and **animal husbandry** contribute significantly to the emission of GHGs; their interaction with the climate is beginning to be better understood.

A systemic analysis is essential!

"A quarter of the carbon footprint of the French comes from their plate," reads the headline of *Les Échos* of February 28, 2019, based on a study by Ademe (French Environment and Energy Management Agency). The French should eat less meat! Ruminants are a source of methane and fertilizers a source of nitrogen oxides, not to mention transportation and agri-food industries, sources of waste, including liquid manure, and polluted water. And what about food waste! As for organic crops, they would be more polluting, from a GHG point of view, than conventional crops, because their yields are much lower. A strange paradox!

During the world climate conference (COP21) held in Paris in 2015, the countries of the world made a commitment, for the first time, to limit global warming below 2 °C – or even less than 1.5 °C, if possible – by 2100, compared to the pre-industrial situation. Simplifying to the extreme, highly industrialized countries must favor a **low-carbon economy**, and thus turn to renewable energies instead of fossil fuels. Will the big polluters keep their word? Will they go towards colossal investments and economic slowdowns?

Climatosceptics (some call themselves **climate realists**) refute the aforementioned assertions. Planet Earth has been through worse during its four billion years of existence, they say. They question the fact that global warming is of anthropogenic origin. For some, it is the sun that is the cause. Therefore there is extensive debate!

The explosion of the Indonesian volcano Krakatoa (or Krakatau) in August 1883 is, without doubt, one of the most violent eruptions ever recorded on Earth. It sent ash over 80 km into the atmosphere and released millions of tons of SO_2. The world temperature dropped somewhere between a few tenths of a degree to and one degree centigrade. It took several years for the climate situation to return to normal. Is it this phenomenon that gave Paul Crutzen the idea of sending sulfur particles into the atmosphere to protect the Earth from cosmic radiation?

2.4. Awareness, sustainable development

2.4.1. *Rachel Carson and sustainability*

After the Second World War, the United States used stocks of an organochlorine pesticide called DDT (dichlorodiphenyltrichloroethane) to protect crops from certain insects. It was sprayed indiscriminately in the air; many birds were killed. This product is one of the "miracle" products of the Second World War: it saved millions of lives from typhus, malaria, and other pandemics. Typhus killed off more of Napoleon's soldiers than the bayonet or the cannon. Some historians believe that it was more responsible for the disaster in the Russian campaign than General Winter.

In her 1962 book *Silent Spring*, Rachel Carson (see Figure 2.7) condemned the uncontrolled use of toxic chemicals. This resoundingly successful book followed on from her 1951 bestseller, *The Sea Around Us* (Dal Pont 2012, p. 31).

This marine biologist showed, in particular, that time must be allowed for the reproduction of aquatic species, which cannot be fished without restrictions and limitations. The term **sustainability** was coined; it is the basis of the concept of sustainable development.

Figure 2.7. *Rachel Carson (1907–1964), inventor of sustainability*

2.4.2. *Sustainable development*

The concept of sustainable development (Dal Pont 2012, pp. 29–30) was highlighted in the 1987 report by Gro Harlem Brundtland (*Our Common Future*), published under the aegis of the United Nations World Commission on Environment and Development. One sentence sums it up: "development that meets the needs of the present without compromising the ability of future generations to meet their own needs." The French translation of "sustainable development" fluctuates between *soutenable* (sustainable) and *durable* (long-term), which is not without ambiguity.

Figures 2.8 and 2.9 try to give a better understanding of this concept and show its practical aspects.

Sustainable development is well represented by the 3 Ps: People, Profit, Planet. It is a development that is:

– socially desirable and accepted by the majority of citizens;

– ecologically sustainable, the impact on the environment of which is under control;

– economically viable:

- point O for "objective" in Figure 2.9 represents the best compromise for a given location, at a determined time. It is defined in metrics;

- Professor Catherine Azzaro-Pantel reviewed a number of metrics in the works referenced (Dal Pont 2012) and (Dal Pont and Azzaro-Pantel 2014);

- without metrics, we very quickly fall into a kind of naïve idealism where, in order to please everyone, we say that the impact of a certain activity on the environment is negligible and that it makes money. This is not the real world, at least not that of the industrial enterprise;

- we understand all the difficulties there are in implementing the concept of sustainable development when we are dealing with the water-energy-food-climate nexus.

For example, the concepts of sustainable agriculture are difficult to define and even more difficult to implement. Thus, just for the wine sector, we should reduce the consumption of water, which is on average 3 liters of water per liter of wine, and reduce it to one liter. This is what the OIV (International Organization of Vine and Wine) recommends, which advocates sustainable vitiviniculture.

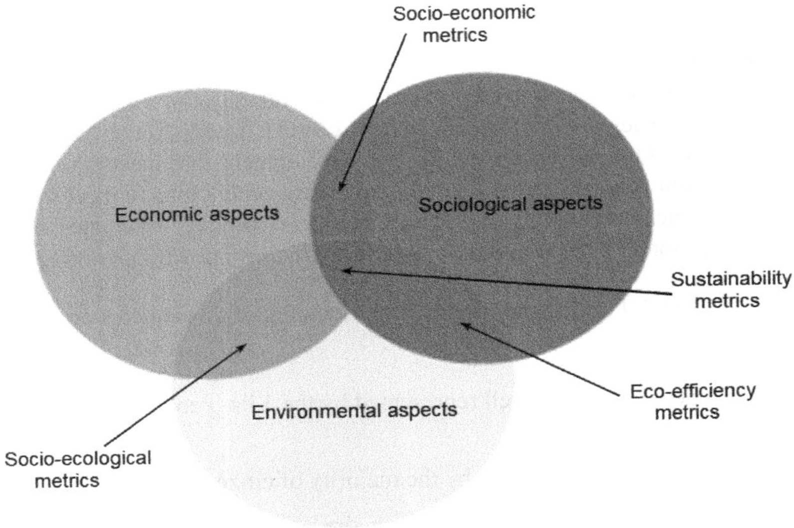

Figure 2.8. *Conceptualization of sustainable development. For a color version of this figure, see www.iste.co.uk/dalpont/process1.zip*

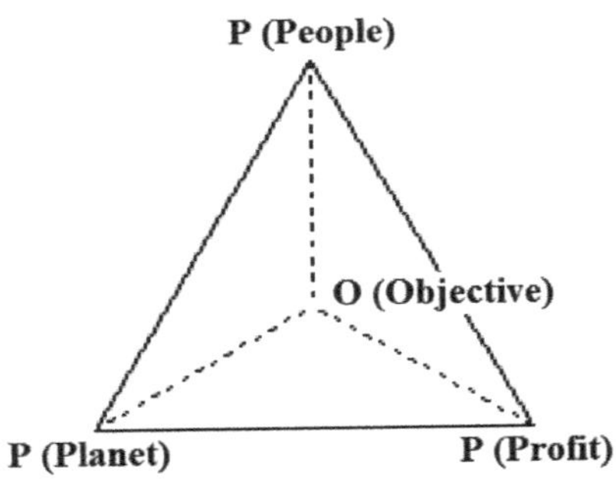

Figure 2.9. *Sustainability and metrics*

2.5. Products by the millions

Products are the essence of any commercial enterprise. They are the ones that make the enterprise live and, for many, represent it: the standard-bearers, we might say. Who does not know Mercedes, Coca-Cola, Champagne, Nylon, and Dior perfumes? The concept of service is often inseparable from the concept of product.

The product is part of the company's strategy, as defined in Chapter 1. The product/market couple became predominant in marketing in the 1980s. There are multiple markets: chemistry, pharmacy, cosmetics, food, wine and spirits, luxury, clothing, sports, household appliances, audiovisual and tourism. Each commercial sector has its niches.

Take the case of automobiles, one of the most important markets: the manufacturer has the choice of selling small or large, gas, diesel or electric cars, with two or four doors, with manual or automatic speed change. It is the role of marketing, the essential function of which is to know the needs of the market, to define the models, which can also vary depending on the country.

Chemistry has long distinguished **commodities** from **performance products (specialities)**:

– commodities are products sold in a catalog with a specification sheet in large tonnage. These are generally products containing a single molecule, not formulated or poorly formulated (see below). These include sulfuric acid, soda, ethylene;

– performance products are for the most part very specific, formulated products, which bring significant added value to the customer.

As an example, we can cite Rhône-Poulenc, which sold lubricants used in the manufacture of cables for radial tires, products whose fine-tuning to suit customers had required years of tests, since the cable is an element essential in tire safety.

The concepts of **function** and **added value** are inseparable today from the concept of the product. When the United States went to war against Japan after the attack on Pearl Harbor on December 7, 1941, it lacked certain products from Asia-Pacific. This was the case of General Electric (GE) where Lawrence D. Miles, who was responsible for finding substitutes, had an ingenious idea. A product was no longer considered as an assembly of parts but intended to fulfill one or more **functions**. The product had to bring **value** to the customer (Dal Pont 2012, p. 317 *et seq.*). This simple idea led to a real revolution at GE; products that came out of factories often cost less and were of better quality. The idea took twenty years to reach France.

Figures 2.10a and 2.10b represent the product as seen by the customer and as seen by the manufacturer, respectively.

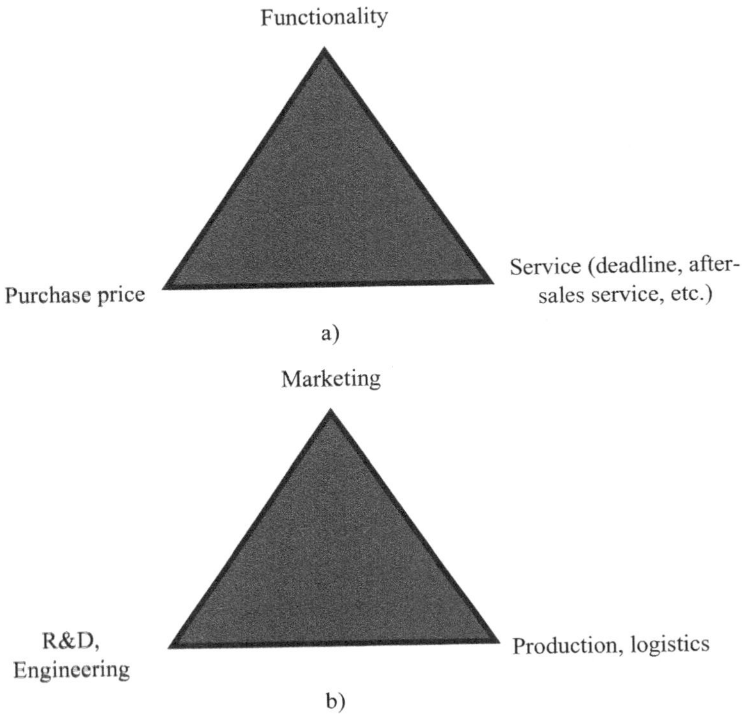

Figure 2.10. *a) The product as seen by the customer; b) the product seen by the company*

Customers look for one or more functionalities; they wish to pay as little as possible, or at least a reasonable price, and hope that their supplier will help them to start using the purchased product. Reliability of supply and consistency in quality are factors that create trust, which is the very basis of commercial success.

Success for the supplier is different in nature. Marketing, the primary function of which is market research, must find new products that add value to a targeted clientele.

The process is a *sine qua non* for commercial success; without a good process, there is no salvation. An industrialist is always at the mercy of a competitor who, for the same product, finds a process with better efficiency, with cheaper raw materials, and who uses fewer steps for its production. Therefore, research is key; the rest will follow "like the cavalry", as a famous French military saying goes. What follows is industrialization and production, which are the subject of the following chapters. Commercial success depends on this **value chain**.

A satisfied customer is a return customer; this is not news. There is nothing worse for a customer than to run out of a product as a result of breakdowns, strikes, and malfunctions of any kind faced by the supplier. This customer then finds themselves facing their own customers who are only eager to find another source of supply.

The customer is *king*! And it is so more than ever, given the current heightened competition, a product of globalization. China has become the "workshop" of the world and **its largest market.** Advertising, the ups and downs of a product, everything is known at the speed of the Internet. The customer/supplier connection has never been stronger, as products are increasingly high-end technological artifacts for progressively sophisticated applications.

Suppliers need to understand the "secret" needs of their customers in order to better serve them, protect themselves from disappointments, and be "first on the market". This has, as we will see, a significant impact on the industrialization process and leads to multiple cooperation, JVs, company purchases, start-ups, and outsourcing of research.

What about planet Earth in all this?

2.6. Resource Earth, garbage Earth: towards a circular economy

This unbridled schumpeterism, which leads to the **creative destruction** of old products by new ones or entire economic sectors by others considered more innovative, creates this profusion of products that are supposed to attract customers fond of new products. This is the case of the smartphone: we want the latest one! Kodak has had a hard time moving from film to digital. There are many examples, and right now, in digital societies, one has to find the right technology, the right niche.

The impact on the planet is illustrated in Figure 2.11.

It has now been established that if the world average standard of living were that of the United States it would need the resources of five Earths – three Earths if it were that of France! We eat our "seed corn", and moreover, our planet becomes a dump. An awareness in line with sustainable development has led to the concept of a **circular economy**, illustrated by Figure 2.11, with the basic realization that our resources are finite: the Earth is a **finite space**!

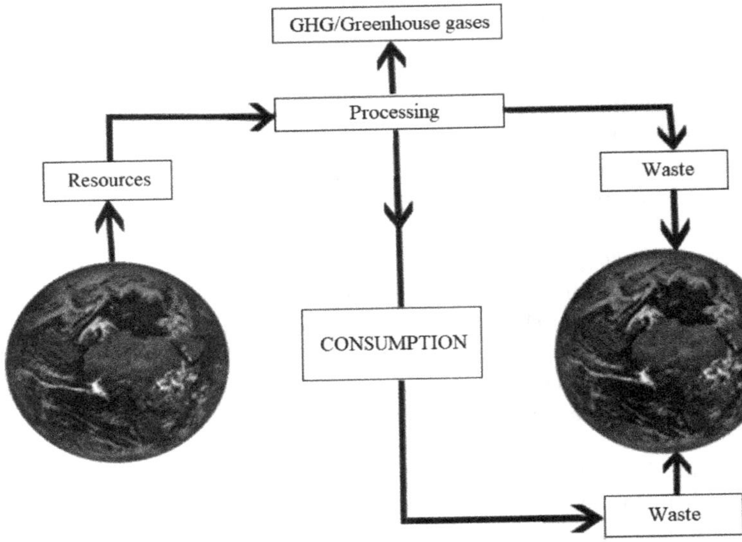

Figure 2.11. *Resource Earth, garbage Earth. For a color version of this figure, see www.iste.co.uk/dalpont/process1.zip*

2.6.1. *Circular economy*

This concept aims to encourage us to emerge from a so-called linear economy that consists of **extracting, producing, consuming, and discarding**. The aim is for products to be reused, and for their waste to be used in the production of new products or to be valued for their energy content.

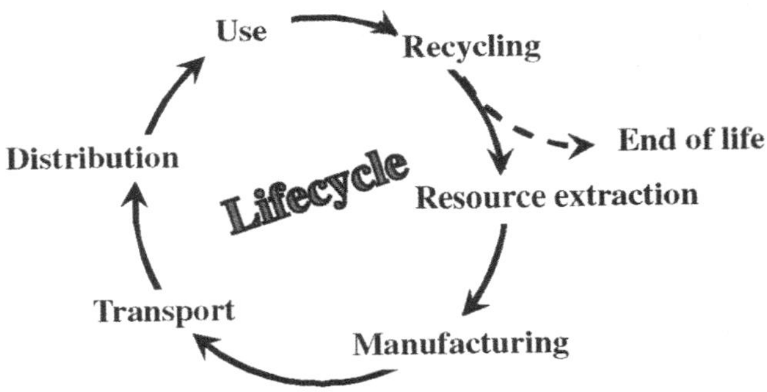

Figure 2.12. *Lifecycle of a product (Dal Pont 2012, p. 244)*

Box 2.5 is typical of this thinking. What could be more common than plastics? What could be more used in human food, clothing, construction, and furniture? They are now victims of their success, ease of use, and low cost. Collecting and sorting them requires specialized companies (see Box 2.5). Identifying and **tracing** them requires increasingly sophisticated sensors, as echoed in Boxes 2.4 (in this volume) and 5.4 (in the second volume).

The above reflection was taken forward and gave rise to the concepts of **LCA** (lifecycle assessment) and **ecodesign**.

2.6.2. *Lifecycle assessment (LCA) and ecodesign*

Ecodesign is a multi-criteria preventive process that seeks to identify and reduce all environmental impact of a product at source. It relies on a powerful tool for identifying environmental impact: lifecycle assessment, or LCA (Sylvain Caillol in (Dal Pont 2012, p. 243)).

LCA, also known as eco-balance, establishes the environmental profile of a product, process, or system for all stages of its lifecycle: from the extraction of raw materials to the product's end of life, which may be recycling, landfill, or incineration. LCA identifies areas of improvement to reduce resource consumption and pollutant emissions; it is a source of progress.

LCA becomes an essential tool in sustainable development approaches. It is focused on assessing the environmental impact of products, services, processes, and businesses. Box 2.2 deals in detail with LCA.

LCA is a method based on ISO 14040 and ISO 14044. It usually includes the following phases[7]:

– defining the objective and scope of the study (functional analysis of the product, determining the system's boundary, etc.);

– inventory of incoming and outbound material and energy flows;

– environmental impact assessment.

Each step results in phases of interpretation to propose ways to improve the system being studied.

The use of LCA is not as simple as it appears and requires a good dose of expertise. Sylvain Caillol (already cited) detailed the study on the LCA of supermarket

7 Source: Ademe, very simplified.

bags from different raw materials: polyethylene or paper, single-use or reusable. The choice of indicators is not easy and the habits of customers must be integrated. It requires relevant bases with **metrics** suitable to the subject studied.

Thus, the so-called electric car, which operates with batteries weighing several hundred kilos seems "ecological", that is to say non-polluting, in the eyes of the citizen. The first question that arises is the origin of its energy: decarbonized, from renewable or nuclear energy, or carbon, of fossil origin. In the latter case, pollution can be considered to be transferred, as CO_2 is generated elsewhere, for example, in the coal-fired power plant. There is also the problem of CO_2 inherent in the automobile manufacturing process, including batteries, and the end-of-life problem of batteries, given that they generally contain toxic materials. We will return to the complex subject of material toxicity later.

Buildings are the subject of specialized LCA studies. This is easy to understand when we know that buildings consume more than 40% of the energy of a country like France and represent 40 million tons of rubble annually (construction and destruction). The choice of building materials is therefore of paramount importance.

2.7. Materials science

That which we commonly call materials, from which the term "materials science" originates, covers an impressive number of sciences, techniques and products, involved in multiple sectors or activities. It is a huge field of activity that is constantly evolving. For illustrative purposes only, let look at a few products, techniques and sectors:

– **energy**: storage, fuel cells, batteries, photovoltaic panels, hydrogen production;

– **anticorrosion products**;

– **surface treatment**;

– **particulate systems, powders**, production (crystallization, atomization, compaction, etc.), handling, storage, packaging;

– **tribology:** the science that studies phenomena that can occur between two surfaces in relative motion; includes lubrication, friction, and wear;

– **plastic materials**: thermoplastic, thermosetting;

– **steels, alloys, ceramics, graphite, glass wool, composite materials**, etc.: aircraft, including Airbuses, are now built with more than 50% composite materials, which are gradually replacing aluminum. They give lightness and improved rigidity. A composite contains at least two immiscible products. For aeronautics, these are usually carbon fibers (reinforcement) with a binder or matrix, often thermosetting resins. In addition to aeronautics, these materials are increasingly applied in shipbuilding, wind energy, automotive industry, construction and civil engineering, sporting goods, cable cars, etc. Because of their nature, they are difficult to recycle, which is not in line with the zeitgeist of today. Research turns to bio-based materials such as hemp, wood, etc.;

– **communication/audiovisual media**: PCs, smartphones, Walkmans, CDs, televisions, cameras;

– **health**: prostheses (dentures, hearing aids, prostheses for hip, knee, limbs, etc.), medical devices;

– **membranes**: health, water treatment (see Box 6.1 of the second volume);

– **agri-food industries**.

2.8. Product formulation and engineering

A chemical product is rarely sold as is; the main product, called an "active substance", is usually formulated with other products that will guarantee its useful properties and one or more features. Virtually all sectors use formulation: chemistry, pharmacy, cosmetics and perfumery, phytosanitation (plant protection), agri-food industries, cleaning products.

A high-end shampoo can contain about 20 different products to perform the following (non-restrictive) functions:

– easy to rinse;

– easy combing;

– drying time;

– shine;

– reducing static electricity;

– soft to the touch;

– volume;

– etc.

The supplier-user relationship is not always simple. Suppliers of surfactants that a cosmetician will seek for shampoos probably does not have the "hair science" that their clients possess, who will use tests of their own that they fiercely keep secret.

Without going any further in this subject specific to each sector, let us note the importance of sampling in testing protocols.

Pharmaceutical formulation allows the active ingredient of a drug to perform the expected curative function using excipients. Chapter 3 and Box 2.7 broadly deal with this field, which has become a true science. It is a branch of chemical engineering in that the formulation implements one or more unit operations: grinding, mixing, compacting, etc. It concerns making the product usable.

2.9. Product toxicology and ecotoxicology

Box 2.3 deals with this thorny and, at the very least, difficult aspect of defining the impact of products, in the broadest sense, on humans and the environment: chemicals, medicines, food products, etc. This has become a major concern of citizens and consumers of developed countries.

Toxicology and ecotoxicology are the disciplines that are the foundations of impact assessment, using standardized procedures, on living organisms and terrestrial and aquatic ecosystems, respectively.

There is often confusion between hazard and risk. A substance can be inherently hazardous. Sulfuric acid (also known as oil of vitriol) is hazardous. It poses risks depending on how it is used and the quantities involved:

risk = hazard × exposure (duration × dose)

The concepts of acceptable levels and doses considered to have no significant toxic effects will lead to the classification and labeling of substances, as well as the definition of preventive and protective measures to be implemented when they are used.

Studies, as described in Box 2.3, are complex: the transformation of products into metabolites (metabolism) sometimes more toxic than the original molecule, their varied uses by people of varying levels of sensitivity (pregnant women smokers, alcoholics) or people already working in an already polluted environment, makes it difficult to analyze the experimental results.

Public opinion, which often lacks education, comes to suspect the information given to it and sometimes reacts violently. The notion of benefit versus negative effect is not well understood. This has already been highlighted for DDT. **A drug is supposed to do good, but it almost always has side effects.** It is important to know how to choose.

The subject of "endocrine disruptors" is of increasing concern to consumers because of its very nature. Alain Lombard described the difficulties of approaches and evaluation.

2.10. Product packaging and ergonomics

Packaging encompasses two concepts: packaging and packing/wrapping.

A formulated product needs to be packaged and wrapped for distribution and placing on the market. This is a considerable area that affects marketing, since the product must appeal to the consumer; logistics; supply chain in general; and therefore carriers, distributors, resellers, and retailers. It's about managing physical and financial flows. We can note the importance of *traceability* and the *blockchain*.

2.10.1. *Packaging and packing/wrapping*

Packaging is the first envelope of the product; it is what contains the product. The packing/wrapping is what makes it possible to transport the wrapped product safely, ensuring easy handling. A drink can be packaged in glass bottles, which will be packed six per carton; the cartons will be placed on wrapped pallets that are easy to handle with forklifts. The bottle has to arrive intact to consumers and correspond to what they ordered.

The packaging must provide information on the product (traceability, identification), guarantee its protection, and maintain its functionality, particularly for food (expiration date). The packaging is sometimes linked to the brand and inseparable from the product: this is the case with perfumes.

We will limit ourselves to these few notions as the domain is immense and riveting. The ergonomics of the product, because of its importance, requires some explanation.

2.10.2. *Ergonomics*

Ergonomics was defined, in 1988, by the French-speaking ergonomics society, as "the implementation of sciatic knowledge relating to man, and necessary for designing tools, machines, and devices that can be used with maximum comfort, safety, and efficiency for the highest number of people" (Daniellou and Naël 1995).

Ergonomics has two fields of action: mass-produced products (food, pharmaceuticals, cosmetics, etc.) and industrial systems. We will focus very briefly on the first category and return to the second in Chapter 3 of the second volume.

Product design, particularly commonly-used products, must take into account the following ergonomic aspects: the ease of gripping and opening food products by the elderly or disabled, and the inviolability and difficulty of opening by children. The packaging of hazardous products, and, particularly, of certain pesticides for agricultural use, must be designed so that their handling does not present an unacceptable risk for the user. The risk posed by polluted packaging must be taken into account.

Our environment includes more and more wrapping material that pollutes both visually and physically. The concept of the 3 Rs (reduce, reuse, recycle) is fully applicable to this invasive area. A "circular economy" approach is required. Sylvain Martin's article (Martin 2019) took stock of the European jungle of standards and directives on the subject. It is thought-provoking.

2.11. New consumer requirements

Consumers in developed countries want to know what they have on their plate, in their glass and the origin of what they consume; they want to know how the product was made. The slogan "from farm to table" reflects a new state of mind in the food sector.

Consumers want a natural product and increasingly want "organic" products; the term "bio-based" is trending. The term "biofuel", denoting a product derived from natural or recycled products (see Box 2.6), is favored by automobile users and, one might say, eases their conscience.

Clearly, the concept of sustainable development has gained ground. So has the concept of greenhouse gases. What appears today as highly-publicized climate change makes citizens of developed countries think.

Naturalness, traceability, and authentication must be expressed in concrete terms for industrialists, who risk suffering media setbacks if they do not comply. New applications allow consumers to make bar codes "talk". Is this the bright side of the digital revolution? Well, why not?

2.12. Boxes

What is a Smart City?

A smart, connected, and sustainable city is an innovative city that uses information and communication technologies (ICT), as well as other means, to improve the quality of life, efficiency of city management and its services, and competitiveness, all while respecting the needs of current and future generations in the economic, social, environmental, cultural, and heritage domains.

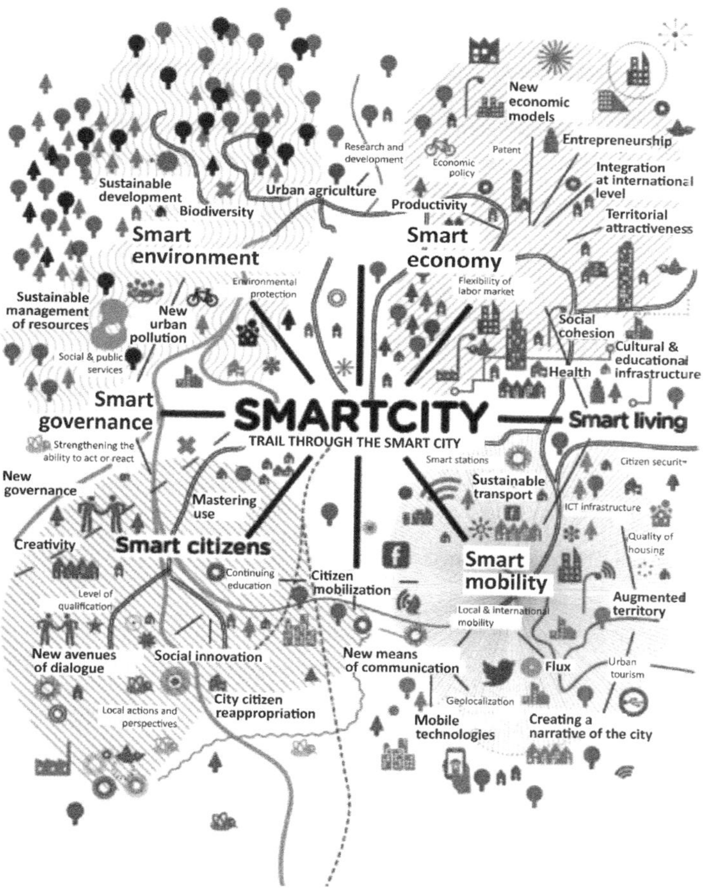

Figure 2.13. *Example of a Smart City. For a color version of this figure, see www.iste.co.uk/dalpont/process1.zip*

Smart City, a multidimensional reality, interconnectivity

A Smart City is a city where "intelligence" is added and the sustainable and interconnected infrastructure of which improves the comfort of living for citizens.

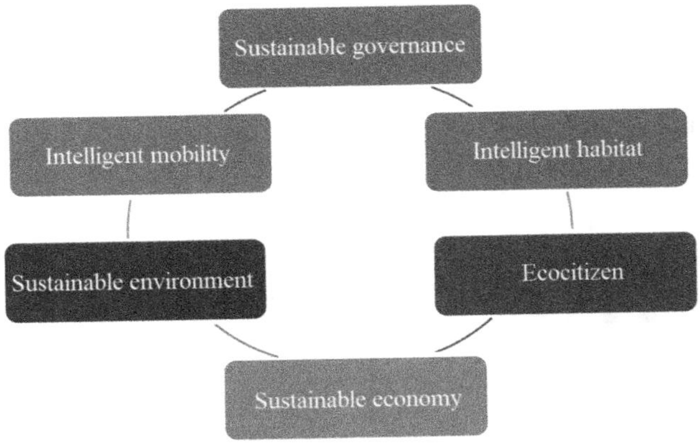

Figure 2.14. *Interconnectivity. For a color version of this figure, see www.iste.co.uk/dalpont/process1.zip*

The six dimensions of a Smart City

Smart economy

– Providing new services, producing new products, and developing new business models.

– Development of the interconnection between the local environment and the globalized environment.

– Development of smart clusters and ecosystems.

Smart environment

– Development of sustainable and green urban planning within cities (roofs, walls, streets, parks, etc.).

– Suitable urban services: adjustable street lighting, waste management, smart furniture, etc.

– Balanced management of natural and heritage resources, reduction of pollution.

– Development of "smart" energy management.

Smart mobility

– Implementation of sustainable, integrated, communicating, and interconnected transport systems (trams, buses, trains, subways, self-driving vehicles, bicycles, scooters, etc.).

– Use of information in real time.

Smart people

Connecting people (smartphones, etc.) and sharing skills, access to education and training.

Smart governance

– Integration of private, civil, and public organizations (municipalities, clusters of municipalities).

– Participatory decision-making (surveys, etc.).

– Access to data, transparency, e-governance, e-services, and new applications.

Smart living

– Lifestyles, consumption, housing, social cohesion, access to culture.

– Quality of life: health, housing, tourism, leisure.

"Citizen centric" services

– Accessing everyday services in the urban space.

– Becoming more flexible in a changing environment.

– Adapting to one's lifestyle, movements and constraints.

– Communicating with others (social).

A comprehensive service offer

Figure 2.15. *The consumer and related services. For a color version of this figure, see www.iste.co.uk/dalpont/process1.zip*

Big Data at the heart of the Smart City

A connected city makes it possible to collect a multitude of data.

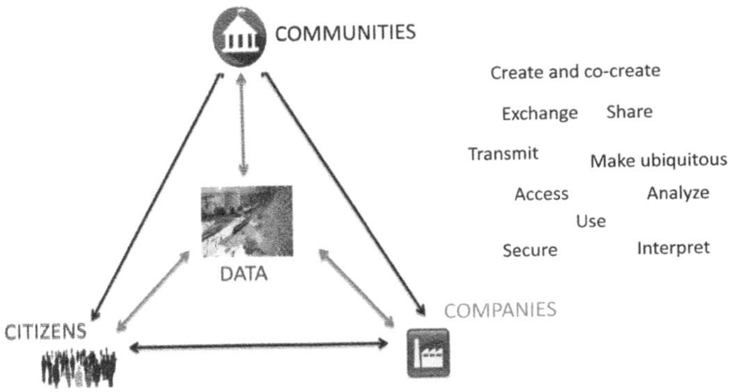

Figure 2.16. *Connected city, a view of central data. For a color version of this figure, see www.iste.co.uk/dalpont/process1.zip*

Innovative and effective services that can organize collaborative intelligence

Riddled with data-sensors that are supposed to improve our urban lives, the Smart City is also a fantastic potential surveillance vector, an aspect that should not be forgotten and which is rarely highlighted by city developers.

What role for objects in Smart Cities?

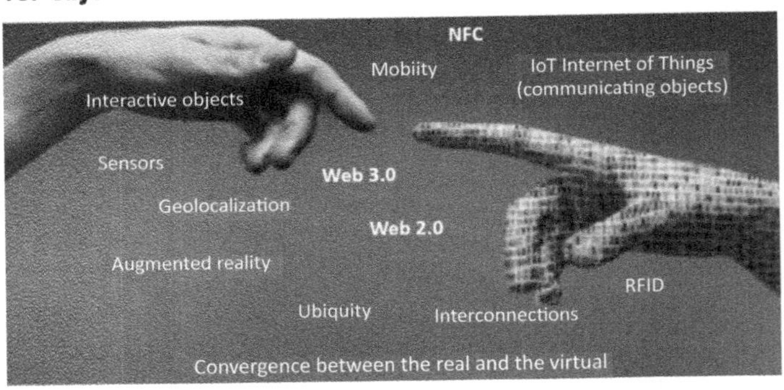

Figure 2.17. *Connected city, a view of central data*

Communicating objects (IoT, Internet of Things) providing the link between the physical world and the digital world will gradually populate the city around its inhabitants (smartphones, location systems, 2D barcodes, QR codes, sensors of all kinds, street furniture holding information, talking objects, communicating objects with or without a screen, carriers of differences in perceptions, etc.).

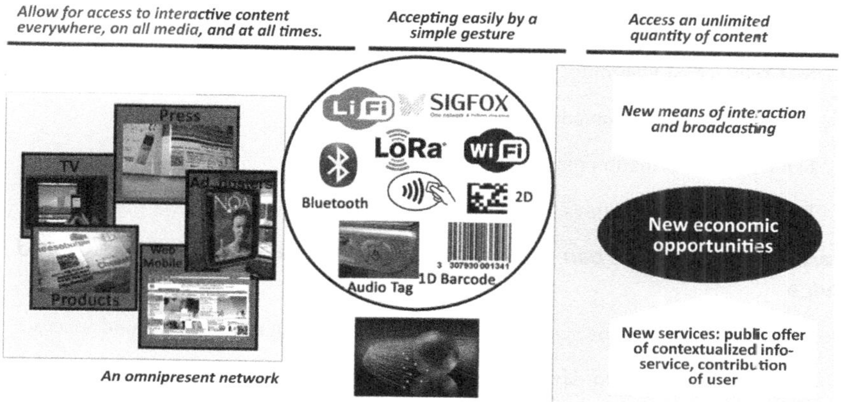

Figure 2.18. *Everyday connected objects. For a color version of this figure, see www.iste.co.uk/dalpont/process1.zip*

App-based standards for connected objects

Most connected objects have a radio link to communicate. Not all connected objects have the same needs in terms of data transfer, range (coverage), etc. For connected objects in the house, there will be Bluetooth, Wi-Fi, Zigbee, Enocean connections, etc. For objects connected on local networks, there will be 5G (Lte and NB-IoT), LoRa, SigFox, or Weightless.

Figure 2.19. *Main standards based on applications (source: microcontrollertips). For a color version of this figure, see www.iste.co.uk/dalpont/process1.zip*

Challenges of communicating objects (IoT)

– Standardization, interoperability.

– Accessibility and permanent connection.

– Self-reliance and learning.

– Listening and acting on one's environment.

– Contextual information management and security.

– Data protection and ethics.

Smart objects in the urban area: public offer of contextualized and shared info service

– Disseminate interconnected networks of objects in urban and underground spaces.

– Augment users' day-to-day environment.

– Provide an open and contributory platform of information and digital resources related to the life of citizens (travel, trade, transport, etc.).

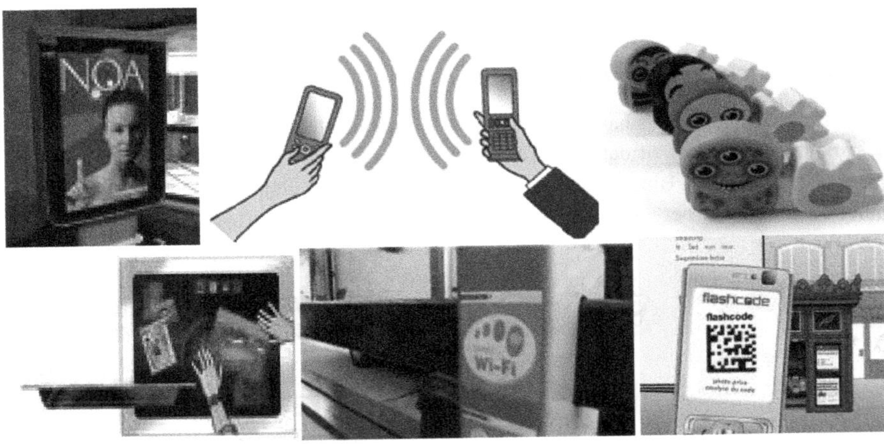

Figure 2.20. *Examples of smart objects in the urban area. For a color version of this figure, see www.iste.co.uk/dalpont/process1.zip*

Communicative street furniture

Ubiquitous daily services, "City Wall", hanging garden, information wells, etc.

Figure 2.21. *Example of communicative street furniture (source: Avestone). For a color version of this figure, see www.iste.co.uk/dalpont/process1.zip*

Other objects in the Smart City: smart public lighting, smart garbage cans, smart benches, reception terminals, vehicle, self-driving vehicles, parking, electric meters (Linky), gas meters (Gazpar), pollution control, etc.

Figure 2.22. *Some objects in a Smart City. For a color version of this figure, see www.iste.co.uk/dalpont/process1.zip*

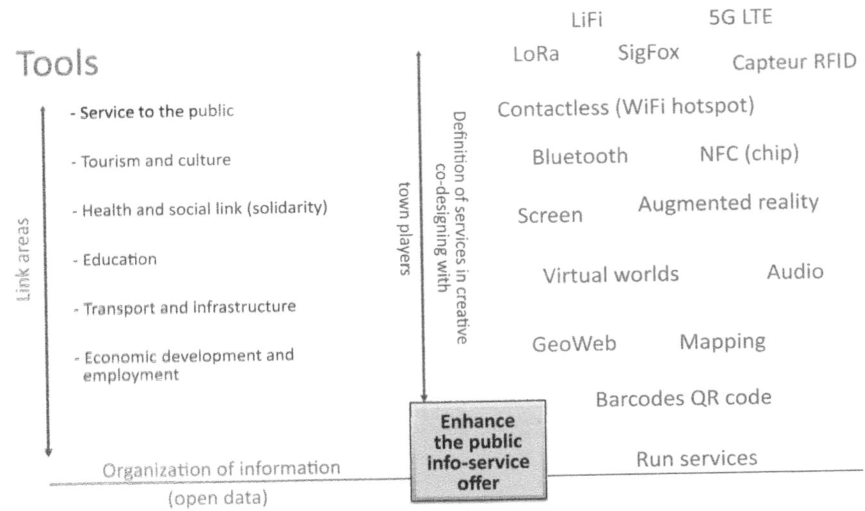

Figure 2.23. *Tools of a Digital City. For a color version of this figure, see www.iste.co.uk/dalpont/process1.zip*

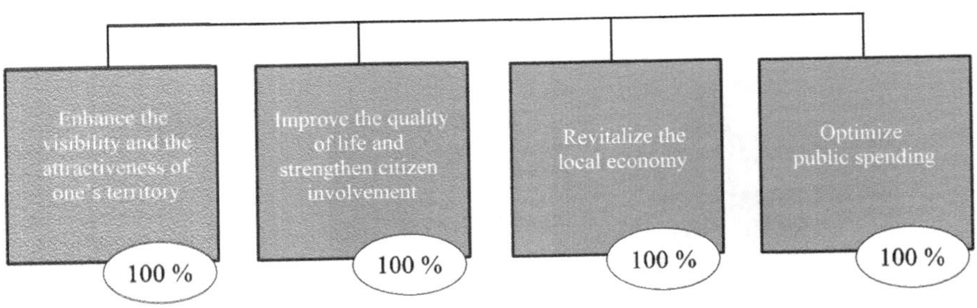

Figure 2.24. *Main challenges of Smart City projects main for the decision-makers interviewed (from: SmartCity Observatory, 2015 Issue, TACTIS)*

Launch of a Smart City

Successfully transforming a city into a Smart City

The possible promoters of the project are:

– a community;

– a cluster of communities;

– a region.

Figure 2.25. *Success factors of a city's transformation. For a color version of this figure, see www.iste.co.uk/dalpont/process1.zip*

Nice is one of the first cities to have embarked on the journey to become a Smart City.

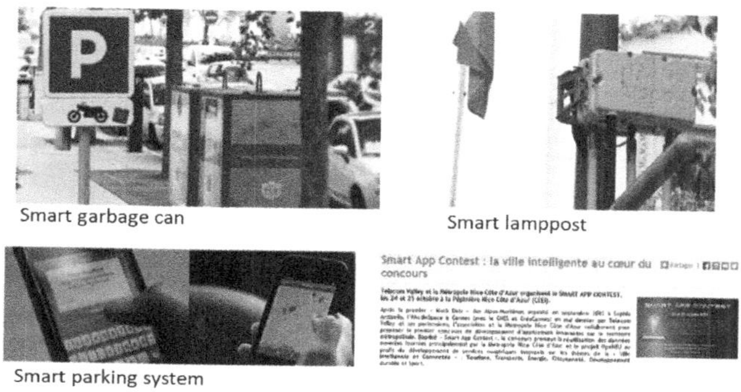

Figure 2.26. *Example of the city of Nice*

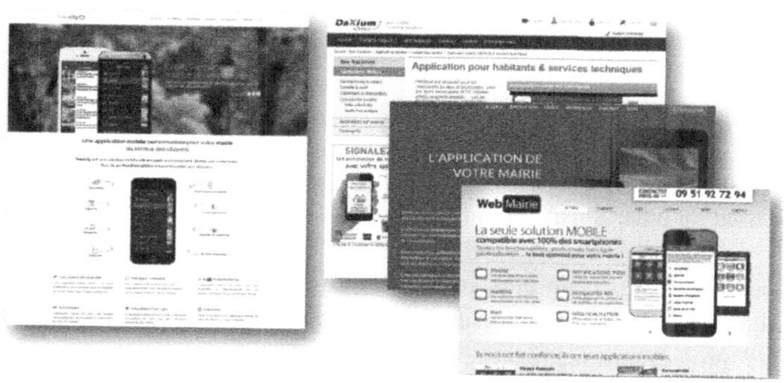

Figure 2.27. *Mobile convergence solutions: local communication, weather, transport, timetables for public services, payment services (parking), and reporting of anomalies (deterioration, waste). For a color version of this figure, see www.iste.co.uk/dalpont/process1.zip*

Smart City: Return on Investment (ROI)

The Smart City offers many benefits to users: quality of life, energy savings, etc.

In financial terms, the return on investment will be long term.

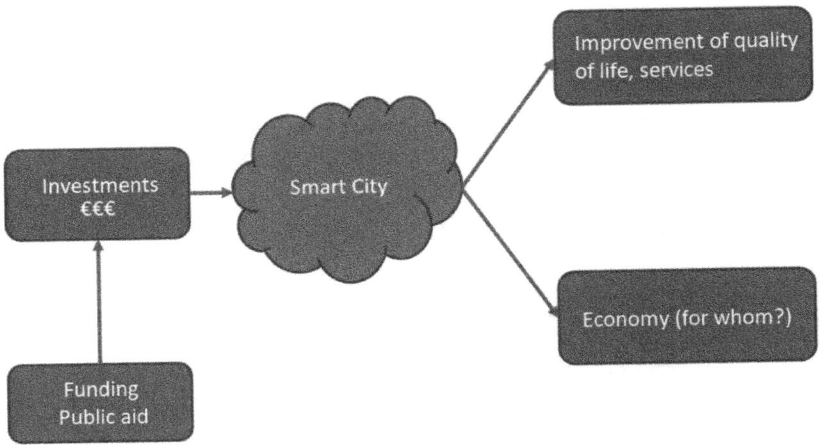

Figure 2.28. *Return on Investment (ROI)*

The Smart City, a sustainable economic model?

In a period of high budgetary constraint, the question of the social and economic utility of public projects is essential, in order to retain only those with high and lasting benefits for territories, local communities and citizens.

Socio-economic assessment makes it possible to quantify the positive and negative, direct and indirect impact of projects and to calculate their socio-economic return. It is a tool for measuring the usefulness of projects.

The socio-economic assessment method has historically been used for transport infrastructure and should be used for other Smart City projects.

Socio-economic assessment thus makes it possible to identify the sources of value creation, accumulating which would make it possible to build new economic models suitable to these "smart" projects or to enrich pre-existing models.

We are only at the initial stage of Smart Cities. The model will evolve according to the needs and development of new technologies, in order to allow for better optimization of costs, resources, and a constant improvement in quality of life.

Philippe Delarue

Born in 1963, Philippe Delarue is an electronic engineer with a specialization in radioengineering. He worked in the development of radar, then GSM mobile phones, and M2M products as the project leader.

He has held various commercial and management positions in start-ups and big companies. Today, he works on connected objects and supports companies in the innovation and new technologies sector.

www.linkedin.com/in/philippedelarue

Box 2.1. *What is a Smart City? (Philippe Delarue)*

Lifecycle assessment for process ecodesign

Integrating the environmental dimension into the technical design of a system is ecodesign. This approach is defined by ISO 14062 (ISO/TR 14062:2002) as a design approach aimed at reducing the environmental impact of products and services over their entire lifecycle, while ensuring identical or improved services to the consumer and the end user. It is a complex process that includes different methodologies and procedures, making it possible to integrate the environmental dimension in the design of a product or a system.

LCA, a quantitative, standardized, and multi-criteria environmental analysis tool

In an ecodesign approach, the environmental impact assessment constitutes a fundamental step and involves different evaluation methods that can be classified into two groups:

– qualitative methods, which aim to rapidly assess the positive or negative effects of an action but without quantifying these effects;

– quantitative approaches, which can themselves be divided into two sub-categories: 1) “environmental” approaches, which quantify the impact of a system on the environment, taking into account all of the “pressures” that it exerts; 2) quantitative “economic” approaches, which economically quantify the effects of a change aimed at reducing environmental impact.

Among quantitative "environmental" methods, lifecycle assessment (LCA) assesses the impact of products, processes, or systems. The purpose of an LCA is to quantify the impact of a product (good, service, or process) over its entire lifecycle, from the extraction of the raw materials that compose it, through the distribution and use phases, and right up to its elimination at the end of its life (from the cradle to the grave). LCA, which has imposed itself today thanks to its standardized nature, is recognized as the most successful method in terms of an overall multi-objective evaluation. It results from the interpretation of the quantified balance of material and energy flows linked to each stage of the lifecycle of the product, process, or system, expressed as potential impact on the environment. An LCA thus involves an inventory step of the exchange of materials and substances throughout the lifecycle, in relation to the functional unit chosen as a reference. The benefit of an LCA is comparing different scenarios from the point of view of environmental impact for the same function of the system.

LCA follows the normative framework ISO 14040:2006 (environmental management – lifecycle assessment – principles and framework) as well as ISO 14044:2006 (environmental management – lifecycle assessment – requirements and guidelines). It has four stages (see Figure 2.29) (Jolliet *et al.* 2017):

– definition of the objectives and scope of the study: determination of the function, the functional unit (spatial and temporal context), and the boundaries of the systems;

– compilation and analyses of an inventory: description of the process, collection and calculation of data, interpretation of the results, and analysis of their limits;

– impact assessment: through classification and then conversion of inventory data into impact (characterization);

– interpretation of results: classification and comparison of scenarios.

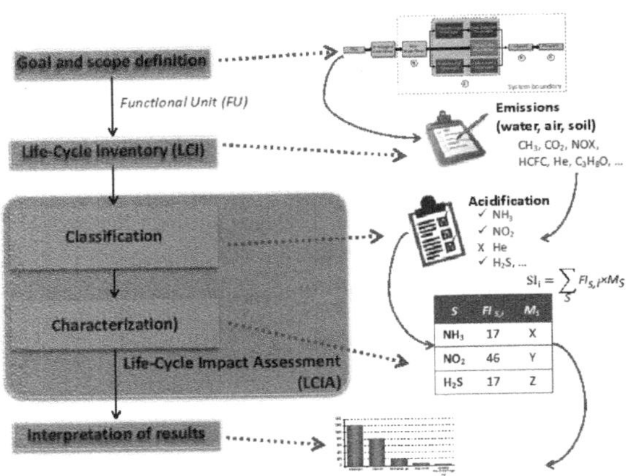

Figure 2.29. *Key steps in an LCA approach (source: Ecoinvent). For a color version of this figure, see www.iste.co.uk/dalpont/process1.zip*

The inventory phase is based on the establishment of input and output flows, which can be carried out through the use of databases (in particular, Ecoinvent) and the use of process simulation software tools (flowsheeting).

An important point of the LCA method concerns the impact method. There are generally two main categories, based on their positioning on the continuum of the cause and effect chain: on the one hand, the so-called "problem-oriented" methods (midpoint), on the other hand, the so-called "damage-oriented methods" (end-point).

The so-called "problem-oriented" or midpoint methods, the most recognized and used today, make it possible for the inventoried flows to be characterized as potential impact indicators (or midpoint indicators) (in the order of tens). They model the impact relatively close to the environmental flow and therefore only concern part of the environmental mechanism. Their advantage is limiting uncertainty (for example, CML 2001 baseline, EDIP97 or 2003).

The so-called damage-oriented or "end-point" methods (for example, EPS and Eco-Indicator 99) model the impact relatively far in the environmental mechanism, that is to say, that which directly damages human health, ecosystems and resources. These indicators are more relevant in terms of communication and are therefore more easily usable, but their modeling is more uncertain due to the complexity of the mechanism and the difficulties in modeling it entirely.

Some methods model impact at both midpoint and end-point levels (for example, the Impact 2002+ method).

LCA and multi-objective optimization in an ecodesign approach

Consideration of the environmental impact by LCA during the process design of a system thus further reinforces the multi-criteria nature of the design problem, also involving technical, economic or social criteria to be satisfied. The problem to be solved can then be formulated as a multi-objective optimization problem and/or a multi-criteria decision aid (Kralisch *et al.* 2015), as soon as conflicting criteria are involved.

Different methodologies have been developed to tackle this problem in chemical engineering. One approach (Figure 2.30) consists of formulating the multi-objective problem by associating environmental objectives with other criteria such as profit maximization, minimization of operating cost, etc. It involves different stages:

– process modeling, generally carried out using process simulators like flowsheeting software: CHEMCAD, Aspen Plus®, HYSYS, ProSimPlus™, etc.;

– evaluation of environmental (by LCA) and techno-economic criteria;

– solving multi-objective problems by deterministic scalarization methods (weighted sum) or ε-limit, or by stochastic procedures like genetic algorithms, which are widely used in many areas of engineering because of their implicit parallelism (Rangaiah and Bonilla-Petriciolet 2013);

– choosing the best compromise between solutions obtained by optimization, for example, by using a multi-criteria decision aid method (AHP, Analytical Hierarchy Process).

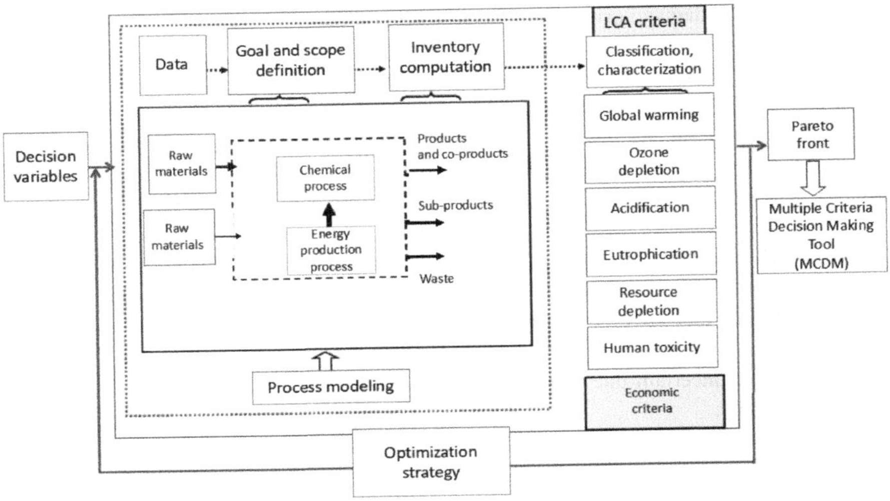

Figure 2.30. *Coupling process modeling, multiobjective optimization, and multiple criteria decision-making*

For more information

Jolliet, O., Saadé-Sbeih, M., Crettaz, P., Jolliet-Gavin, N., Shaked, S. (2017). *Analyse du cycle de vie, comprendre et réaliser un écobilan*. Presses polytechniques et universitaires romandes, Lausanne.

Kralisch, D., Ott, D., Gericke, D. (2015). Rules and benefits of Life Cycle Assessment in green chemical process and synthesis design: A tutorial review. *Green Chem.*, 2(17), 123.

Rangaiah, G.P., Bonilla-Petriciolet, A. (2013). *Multi-Objective Optimization in Chemical Engineering: Developments and Applications*. John Wiley & Sons, New York.

Catherine Azzaro-Pantel

Catherine Azzaro-Pantel received her PhD in Chemical Engineering from the *Institut National Polytechnique* (INP), Toulouse, France. She is a Professor of Process Systems Engineering at the *École Nationale Supérieure des Ingénieurs en Arts Chimiques et Technologiques* (INP-ENSIACET), University of Toulouse, France, where she co-founded a Master's level program in EcoEnergy. This program's goal is to provide engineers with a state-of-the-art education in the area of advanced energy technologies and systems. Her research interests lie in the area of process systems engineering with a specific focus on optimization methods for process design. She is the author or co-author of over a hundred scientific publications, including articles, conference proceedings, a book and several chapters.

catherine.azzaropantel@ensiacet.fr

Box 2.2. *Lifecycle assessment for process ecodesign (Catherine Azzaro-Pantel)*

Managing the impact of chemical substances and products

Notions of toxicology and ecotoxicology

Societies have a need to live in a healthier and unpolluted world, which allows each human, wherever they live, to live in a clean and preserved environment, in order to be able to eat and develop while protecting the planet's resources.

Natural or industrial chemical products and substances, whatever their applications and uses, must meet criteria of resource conservation and recycling, as well as safety for human and animal health and environmental protection.

The potential impact of chemical products and substances is assessed in toxicology and ecotoxicology studies.

Definitions of toxicology and ecotoxicology

In order to ensure the safety of chemical substances, and industrial, pharmaceutical, and phytosanitary products manufactured by industrialists, it is necessary to study their properties and assess their hazards according to standardized procedures, defined by international regulatory authorities.

Toxicology is the science of poisons; it studies the impact of natural or manufactured chemical products/substances on living organisms.

Ecotoxicology, as its name suggests, attempts to combine two very different subjects: ecology and toxicology. Ecotoxicology is the study of the harmful effects of chemical products/substances on terrestrial and aquatic ecosystems.

Concepts of hazard, risk, and risk management

According to the famous Swiss alchemist Paracelsus (1493), "All things are poison, and nothing is without poison; the dosage alone makes it so a thing is not a poison."

– **Hazard** is a **static concept of potentiality**. It is the set of physical and toxicological characteristics posing a potential hazard. It is an intrinsic property that characterizes a chemical product/substance.

– **Risk** is a **dynamic concept of contingency**. It is the set of circumstances that make it possible for a potential hazard to manifest:

risk = hazard × exposure (duration × dose)

NOTE.– This paradigm (risk = hazard × exposure) applies to the majority of substances with the exception of endocrine disruptors, and certain mutagenic and carcinogenic substances, for which exposure – sometimes one-off – at a low dose could be enough to trigger harmful biological effects.

– **Risk management** is a **subjective concept of predictive assessment**. It is the set of means employed to avoid, or limit to acceptable levels, the manifestation of a hazard. Acceptable levels are defined according to toxicological and socio-economic criteria, specific to each country or organization.

This makes it possible to define "doses considered to be without significant toxic effects" under normal conditions of use. This results in the classification and regulatory labeling of these chemical products/substances, according to an international system, in order to inform users of the hazards of these substances, but also to define the appropriate prevention and protection measures to control the risks resulting from potential exposures.

Hazard and risk assessment of substances and products

Toxicology and ecotoxicology studies are carried out on living organisms or systems (cell cultures, laboratory animals, ecosystems) according to international standard procedures that ensure their reliability.

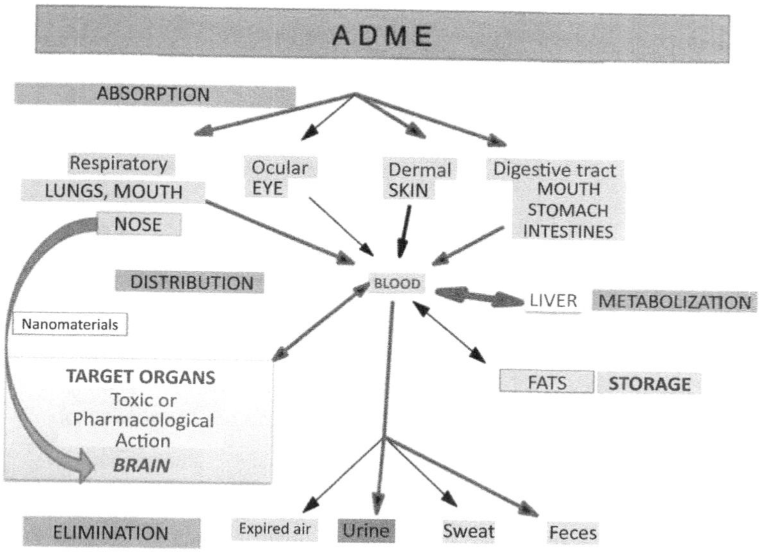

Figure 2.31. *Absorption, distribution, metabolization, and excretion (ADME). For a color version of this figure, see www.iste.co.uk/dalpont/process1.zip*

It is necessary, first of all, to ensure the bioavailability of the substance/product in the organism/ecosystem and to know its absorption-distribution-metabolization-excretion (ADME) outcome. For example, ethylene, classified as non-carcinogenic (also secreted by apples), is transformed in the body into ethylene oxide, which is particularly toxic and classified as carcinogenic. The kinetics of transformation and metabolism in the body is important in the risk linked to ethylene, and to the consumption of apples, which also synthesize ethylene.

The study of the lifecycle and the outcome of substances/products in the environment is important. The physical (radiation, heat), chemical and biological degradation process, and the persistence of a substance/product in the environment, are also evaluated in order to know its bioconcentration, transformation rate (kinetics) and the by-products of degradation (metabolites), the toxicity of which is also evaluated.

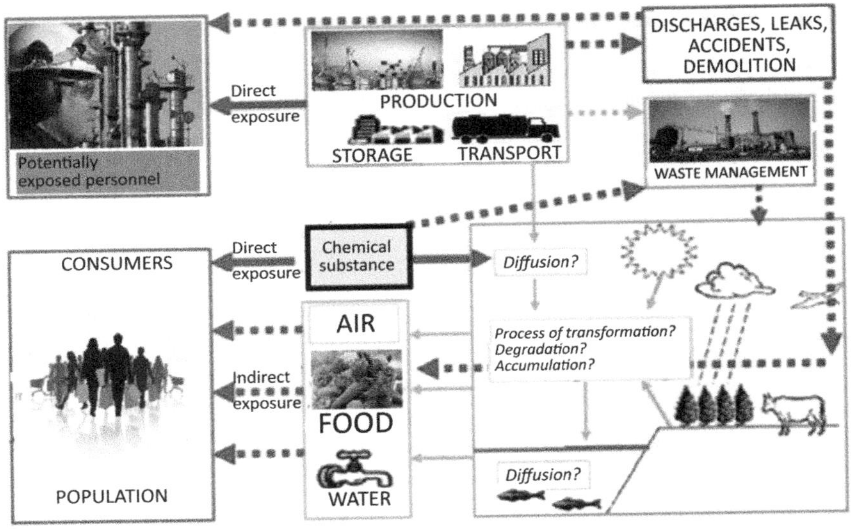

Figure 2.32. *Lifecycle of a chemical substance. For a color version of this figure, see www.iste.co.uk/dalpont/process1.zip*

The study of the impact of chemical products/substances in the aquatic environment is done on various trophic levels of the food chain, in order to study the specific toxicity and also to assess the level of bioconcentration.

At the first level, we find microscopic algae, which at the second level are eaten by micro-crustaceans (daphnia), which at the third level serve as food for fish, which are, at the end of the food chain at the fourth level, eaten by carnivorous fish. Levels three and four are likely to be eaten by humans. The bioconcentration of a pollutant in organisms increases from one level to another and can reach toxic concentrations.

For example, we can cite the Minamata disaster where the dumping of industrial residues containing mercury in the water of a lagoon led to a bioaccumulation of mercury in the fish – which served as food for fishermen – to the point of causing very serious neurotoxic and teratological effects in fishermen and their children.

In order to fully understand the risks of each chemical product/substance, a "toxicological profile" needs to be established by testing each type of organ, organism, or ecosystem by increasing doses that make it possible to define an exposure dose without effect, an exposure dose with minimum reversible effect, and an exposure dose with serious and irreversible effects. Carcinogenic, mutagenic, and reprotoxic (CMR) effects are studied, in particular, in widely used substances and products.

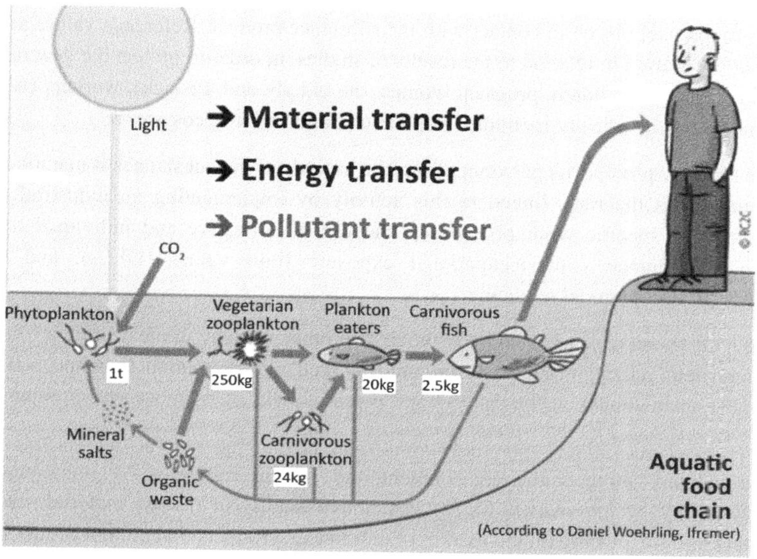

Figure 2.33. *Aquatic food chain and transfer of pollution. For a color version of this figure, see www.iste.co.uk/dalpont/process1.zip*

Endocrine disruptors, acting on a cellular level in a hormonal mode of action, more or less escape these systems of dose definitions since their actions, which are triggered at very low doses, compete with hormones. The same can be true for certain carcinogens and mutagens.

The complexity of the transposition of the cell or animal to humans does not make it possible to detect all the toxic hazards of chemical products/substances which can appear over time, and the multiplication of human exposure. Epidemiological studies make it possible to note deleterious effects of substances/products on humans, but their reliability depends on the parameters observed and especially on the size of the cohort studied for sufficient statistical processing of data. A small cohort (fewer than 1,000 people) does not have sufficient power to establish a causal relationship (exposure/effect) of a substance/product on a population, which explains the controversies over the results of partial or incomplete studies (for example, controversies over the hazardousness and carcinogenicity of glyphosate).

Risk management of chemical products and substances

Based on the information collected and the doses/concentrations observed in toxicology studies (LD50, No-Observed-Effect Level (NOEL), No-Observed-Adverse-Effect Level (NOAEL), etc.) and ecotoxicology studies (No-Observed-Effect Concentration (NOEC), No-Observed-Adverse-Effect Concentration (NOAEC), etc.) on each substance/product, reference values, corresponding to safe levels of exposure, ensure protection in case of exposure for different groups likely to be in contact with the substance/product. Reference values are set by adding a safety margin in relation to the results of studies, in order to protect the general human population (including children, pregnant women, the elderly and the sick), workers (who are a more homogeneous medically-monitored population), and various ecosystems.

– The industry protects its personnel from the hazards of the substances it manufactures or handles and limits the risks linked to this activity by implementing an industrial hygiene policy, including specific work procedures, the use of collective and individual means of protection, compliance with occupational exposure limit values (OELs), and medical surveillance by occupational medicine.

– The industry protects users and consumers and limits risks by implementing a responsible risk management policy, product stewardship as well as information (labeling, safety data sheets (SDS), instructions) on the dangers and potential risks of its substances/products on the market.

– The industry protects the environment and natural resources by implementing an eco-design policy, which ensures the environmental quality of its raw material supply, the manufacture and the placing on the market of substances/products that are not or less harmful, easily biodegradable and not persistent. It studies the lifecycle (LCA) and endeavors to reduce or even eliminate nuisances, by ensuring the possibilities and the quality of recycling or elimination.

– The industry protects the population surrounding its plants against fugitive or accidental emissions of substances/products outside its walls by enforcing a policy for process management, risk calculation (toxicological reference values, thresholds for irreversible effects, thresholds for lethal effects, Acute Exposures Guideline Levels), anticipation, and information of surrounding communities likely to be affected (risk prevention plan).

– The industry protects the environment surrounding its plants against fugitive or accidental emissions of substances and products by implementing a policy for managing gaseous, liquid, or solid emissions in the natural environment, calculating possible risks for the environment, anticipation, and information (risk prevention plan).

In conclusion, knowledge of the toxicology and ecotoxicology of substances and chemicals is an essential factor for the development of the industry of the future. The very early integration of a control and management policy for manufactured substances and products, by knowing their hazards and risks, facilitates the management of the risks of industrial activities and their sustainability.

For more information

Bakès, J., Lemaire, P. (2007). *Écotoxicologie : évaluation des propriétés des substances chimiques*. Techniques de l'Ingénieur, Saint-Denis.

INRS, INERIS. Fiches toxicologiques.

Lombard, A. (2015). *Toxicologie industrielle*. Techniques de l'Ingénieur, Saint-Denis.

Lombard, A. (2018). *Perturbateurs endocriniens. Problématique et perspectives*. Techniques de l'Ingénieur, Saint-Denis.

Techniques de l'Ingénieur. Documentation scientifique et technique [Online]. Available at: https://www.techniquesingenieur.fr/.

Alain Lombard

Doctor of Science, pharmacotoxicologist, and former pharmacology assistant at the Faculty of Medicine in Nancy, Alain Lombard is a toxicologist at Searle R&D in Sophia-Antipolis, CIT in Evreux and Hazleton in Lyon, industrial toxicologist and head of the toxicological industrial hygiene department at Arkema, Paris-La Défense (previously Atofina, Elf Atochem, Orkem). He is also a consultant in occupational risk prevention (IPRP) and in technical and organizational skills (CRAMIF). He is a founding member of the French Occupational Hygienists Society (SOFHYT) and a consultant in industrial toxicology (Allotoxconsulting).

Box 2.3. *Managing the impact of chemical substances and products. Concepts of toxicology and ecotoxicology (Alain Lombard)*

Traceability: principles and applications

Traceability has entered our customs, our habits, our requirements. There is no longer a question of not knowing where products come from and for manufacturers, there is no longer a question of leaving anything unclear in process monitoring, quality and distribution channels.

Traceability is defined as the ability to recover the history of a product using recorded information. This definition says it all, but the effectiveness of a traceability chain depends on the completeness and relevance of the information recorded, the reliability and security of storage, and the sharing of this information to ensure the continuity of the chain. Also, seeing traceability only as a crisis management tool is very simplistic, because the objective must be to avoid the crisis and therefore to react as soon as possible to any process drift.

Tracing begins with identifying with certainty. Whatever the stage in a product's life, from the raw material to recycling, it must be possible to reconstruct the supply, processing, and distribution chain. For this, each input, each semi-finished product or intermediate formulation and each finished product must be identified. Likewise, each actor in the chain must be known.

The multiplicity of internal and external sources and stakeholders required the creation of coding standards to allow the interoperability of systems and thus the sharing of information. Likewise, the creation of tools for automatically capturing identifiers has made information gathering safer and more efficient.

We thus saw the birth of the 1D EAN13 barcode, well known in mass distribution or pharma, and then 1D codes, the best known of which are the Datamatrix and the QR code and RFID/NFC "radiofrequency" codes.

Figure 2.34. *Examples of 1D and 2D pharma codes. Note that since February 9, 2019, the European Falsified Medicines Directive (FMD) requires a single coding (serial number) in the form of Datamatrix. In the above example, only the batch number, the manufacturing date, and the expiration date are coded*

Attached to the identifier, there is a set of information that must be reliable and useful or, is simply mandatory. In fact, the evolution of technologies for measurement and analysis, or even dematerialization, has made it possible to introduce new requirements into regulations and directives that govern traceability in various sectors. Without necessarily specifying the means, the traceability requirement is enshrined in the texts with an underlying requirement for reactivity.

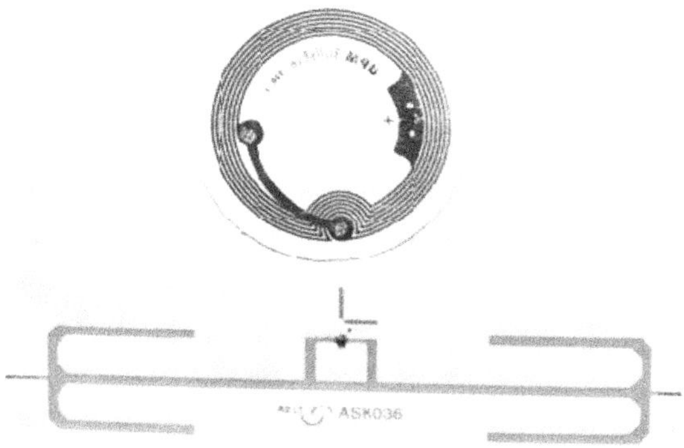

Figure 2.35. *Examples of UHF and HF RFID tags. The UHF tag has a dipole type antenna and operates in the 900 MHz band. The applications are logistical. The HF tag has a wound circular antenna and operates at 13.56 MHz. The applications are logistics and access control*

Identifiers therefore make it possible to capture, aggregate and link product-relevant data. These data are acquired by sensors, which can be simple readers, and which will generate an event (a certain product seen at a certain time in a certain place) or more if said sensor has advanced functionalities (image taking, fingerprinting/signature, real-time analysis, etc.). And these sensors are connected to the company network to store and centralize data internally or externally, if the company uses an external service provider to process its data. A security and data protection problem for the company may arise if a data manager seeks to take advantage of it for personal gain. Likewise, a weakness in network security can open doors to malicious intrusion. There are risks to be weighed up. The French Directorate General for Enterprise (DGE), the French General Directorate for Internal Security (DGSI), and the French Data Protection Supervisory Authority (CNIL) have published notes on the risks of sensors and other connected objects.

Once the data has been collected, the challenge is to also link them together, to constitute an exhaustive and inviolable chain that can be traced back and forth to follow the tree structure. For example, a defect found on one component of a car model will *de facto* create a risk of failure in other models, even if the defect has not been revealed. Likewise, non-compliant meat identified in a cooked dish *de facto* affects other preparations using the same input source.

All of this translates into, but is not limited to, the following needs and constraints:

– rapid analysis, almost in real time, at the manufacturing process level with the implementation of new online sensors and analyzers;

– detection of anomalies as far upstream as possible;

– automation of the collection and aggregation of product-related information and strengthening of data management within the company;

– security and reliability of shared information channels;

– use of the tools of the digital revolution, which is profoundly changing the corporate world: Big Data, Internet of Things, artificial intelligence, real-time management, etc.

Blockchain is promising, we are told, to make traceability more efficient and reliable. Let us try to clarify what is hidden behind this much-used term.

We will leave aside cryptocurrencies like Bitcoin, the native application of the blockchain principle, to focus on the application of the concept on physical products. Cryptocurrency chains aim to guarantee the effective value of transactions carried out. They are, by definition, public channels open to everyone. Conversely, "product" chains are private chains, similar to what some mass food retail chains are currently setting up.

The principle of blockchain is to link together "transactions" that are all milestones in the life of a product. These transactions are packaged in blocks by a "miner" who signs and binds them. There is then no possibility of modifying blocks, or breaking the chain to remove blocks or add others. The chain is therefore safe. And in the blockchain principle, the ledger that contains all the transactions is accessible by all stakeholders. Everyone can therefore check the consistency of the chain.

The only caveat is that, in the blockchain principle, the very content of the information contained in the transaction is not verified by the miner, except of course for cryptocurrencies. This problem can be solved if, as in the case of private and concentrated blockchains, there is only one miner who defines the content of the transactions by standardization, and is able to verify them.

To give an example, the compliance of a product with standards or regulations is recorded in a blockchain under the responsibility of the declarant. If this is fraud, the blockchain secures… fraud!

In conclusion, traceability is made essential by laws, regulations, directives, norms or standards, or simply by good practices that must be observed by any responsible industrialist. Numerous methods, techniques and technologies facilitate the implementation of traceability and compliance with requirements. In view of these requirements, manufacturers must conduct a risk analysis, both with regard to their processes and procedures, as well as with regard to the solutions that they plan to implement.

For more information

– on identification and coding standards: GS1 France www.gs1.fr.

– on blockchain: Blockchain France, www.blockchainfrance.net.

– on connected objects and the Internet of Things: https://en.wikipedia.org/wiki/Internet_of_things.

– on the security of connected objects:

- DGCCRF: https://www.economie.gouv.fr/dgccrf/Publications/Vie-pratique/Fiches-pratiques/objets-connectes;

- CNIL: https://www.cnil.fr/fr/objets-connectes-noubliez-pas-de-les-securiser;

- DGSI: https://www.entreprises.gouv.fr/files/files/directions_services/information-strategique-sisse/publications/FI-49-_janvier-vulnerabilites-induites-utilisation-objets-connectesl.pdf.

Jean-Michel Loubry

An expert in traceability and protection against counterfeiting, Jean-Michel Loubry was convener of the AFNOR GEPPC group of experts from 2008 to 2018, dealing with protection against counterfeiting and forgery. He is also a former director of the National Traceability Center in Valence.

Box 2.4. *Traceability: principles and applications (Jean-Michel Loubry)*

Recycling entrepreneurs at the heart of the circular economy: The special case of plastics

The recycling industry has made waste the 7th resource on our planet. It has become a high-tech industry.

It is one of the responses to the **fight against climate change** to which our society is increasingly attentive, by reducing the consumption of materials, energy and the associated carbon footprint. The pollution of oceans, whose gyres are the “sixth continent”, increasingly moves world opinion.

Federec, in partnership with ADEME, has conducted a study which gives very precise quantification of these resource savings at the French level. Each year, the emission of 22.5 million tons of CO_2 is avoided and 124 TWh is saved, thanks to the collection and recycling of 105 million tons of waste.

The role of recycling enterprises is to transform the waste collected from companies and households into recycled raw materials and to market them by meeting precise specifications and the growing needs of consuming industries, such as foundries, steel mills, paper mills, etc.

Recycling is a complex set of operations, as shown in Figure 2.36 for plastic materials. It is these products that we will focus on in what follows.

Figure 2.36. *Lifecycle (source: Agathe Pernet). For a color version of this figure, see www.iste.co.uk/dalpont/process1.zip*

On average, a French person generates 568 kg/year of household and similar waste (according to ADEME). In terms of household waste, plastic represents 23% of selective collection (i.e. 1,152 thousand tons in 2017). With additions to sorting instructions, the tonnages should increase further.

The French law on energy transition for green growth in 2015 aims to reduce waste storage by 50% by 2025. In 2018, the French Prime Minister published the "circular economy" road map (FREC), confirming the policy decreed by the President of the Republic in the summer of 2017 that wants 100% of plastics to be recycled by 2025 and raw materials from recycling to be used as a priority. At the end of 2018, the European Union validated the "single use plastic" directive that obliges Member States to collect 90% of plastic bottles by 2029 and incorporate recycled materials by up to 30% in plastic bottles by 2030. On the basis of all of these texts, France is currently preparing a "circular economy" bill.

To meet these objectives, three directions must be pursued:

– R&D and innovation must include **eco-design for recycling** in the specifications of new products in the same way as the functionalities, that is to say the use values. The products must be **eco-designed**;

– waste collection must be improved and create economic conditions such that they are primarily oriented towards recycling, rather than incineration or landfill;

– more raw materials from recycling must be incorporated into the new so-called "virgin" products through economic incentive mechanisms.

The European plastics business amounts to 27 billion euros for 60 million tons, and represents 1.5 million **non-relocatable** jobs, including around 30,000 in France. Today, only 15% of plastics are incinerated, 14% recycled, the rest ends up in storage, and a minor part ends up in the wild.

The figures for France are as follows:

– recycling: 22%;

– energy recovery: 43.5%;

– landfill: 34.3%.

These figures show the extent of the progress to be made in terms of research and development and in terms of public information and education, without forgetting the industrial and organizational aspects of a leading industry.

The most frequently used plastics are PET (polyethylene terephthalate for bottles), high and low density polyethylene, polystyrene, polypropylene (PP), PVC, polyacrylates, etc.

Table 2.2 shows the applications of virgin plastics.

Name of plastic	Acronym	Examples of use
Polyethylene terephthalate	PET	Transparent bottles
High density polyethylene	HDPE	Opaque bottles
Low density polyethylene	LDPE	Bags, trays, films
Polystyrene	PS	Plastic tableware
Polypropylene	PP	Food containers, bottle caps, chips packets
Polyvinyl chloride	PVC	Windows

Table 2.2. *Applications of virgin plastics*

Recycling companies are adapting to recycle more and better, in order to guarantee a circular economy.

Processing operations call on chemical engineering techniques (handling, sorting, grinding, washing, reaction). They use increasingly sophisticated equipment.

Upstream companies (bottling, packaging in general) must take into account the downstream. A systemic vision of this industry is therefore essential.

Recycling enterprises in the era of digital transition

The recycling industry is strongly influenced by the fourth industrial revolution. The dematerialization of services continues at high speed. Consumers are increasingly buying online and using home delivery services. This has led to a very significant increase in cardboard consumption worldwide, which has had the effect of "boosting" the paper industry. Conversely, the dematerialization of the press has caused a continuous decrease in paper consumption.

Waste marking and the growing use of a wide variety of sensors make it possible to differentiate waste on sorting belts and promote their reuse. Robotics, cobotics and image recognition greatly improve sorting operations and reduce the arduousness of tasks.

The IoT (Internet of Things) steps in to measure and report the level of waste containers. AI (artificial intelligence) optimizes the collection circuits. We are only at the beginning of a real technological and cultural revolution.

The challenges for the coming decades are immense. Each sector has its specificities: plaster, rubble, building glass, floor coverings, etc., require specific solutions. It is becoming necessary to recycle electronic devices, rare earths and precious metals. Chemical recycling makes it possible to go upstream in the chain (to the monomer by dissolution or depolymerization, pyrolysis, gasification). Its advantage is that is makes it possible to ensure the quality of the recycled material for flows which, in present times, are either not or are recycled less for lack of markets. This can also allow for more qualitative applications (food contact, for example). These projects are significantly increasing, but a viable economic model remains to be defined.

For more information

Federec website: http://federec.com/.

Videos
Federec YouTube channel: https://www.youtube.com/channel/UCLbvSrNT4SenVBjWXFtRQ3w

Federec

Founded in 1945, Federec (The Federation of Recycling Enterprises) represents 1,100 companies in the recycling sector in France, with a turnover of 9.05 billion euros. They invest more than 500 million annually, or 6% of turnover.

These very diversified companies range from multinationals to SMEs and ISEs. Spread throughout France, their activity is the collection, sorting, and recovery of industrial and household waste materials or the trading/brokerage of raw materials from recycling.

Jean-Philippe Carpentier

President of Federec, Jean-Philippe Carpentier was elected to the presidency for the first time in December 2012 (3-year term), and was re-elected in December 2015, and then in December 2018.

Box 2.5. *Recycling entrepreneurs at the heart of the circular economy: The special case of plastics (Jean-Philippe Carpentier)*

Biofuels: A response to the energy transition

Energy transition is a major issue of our century. To cope with this, the historic agreement ratified by all the States in Paris during the COP21 (United Nations Climate Change Conference) set an ambitious goal of limiting the rise in global temperatures to a maximum 2°C by 2100. As the transport sector plays an important role in greenhouse gas emissions responsible for global warming (responsible for 23% of energy-related emissions), it is necessary to achieve this objective to develop alternative solutions to fossil energy and to use an energy mix that is more respectful of the environment, particularly with the widespread use of fuels produced from renewable resources.

Since the 1990s, IFPEN and Axens have been developing and offering innovative technologies for the production of biofuels, including Vegan® technology, which allows the production of biojet fuel and/or renewable diesel fuel from lipids.

General Principles of Vegan® Technology

Feed

Vegan® technology makes it possible to produce drop-in fuels, that is to say whose chemical structure is strictly identical to fuels of fossil origin, from molecules that constitute the fat of living beings – lipids.

Lipids are oxygenated molecules present in significant quantities in our environment; they are found in vegetable oils (virgin or used in cooking), algae, animal fats, etc. Their chemical structure (degree of saturation, chain length, composition, impurities) varies from one resource to another.

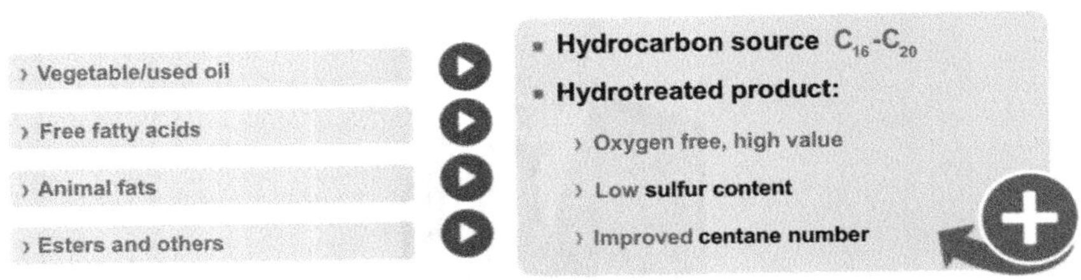

Figure 2.37. *Typical feeds for Vegan® technology*

Technically, any lipid raw material, resulting from the groups of raw materials indicated in Figure 2.37, can be treated with Vegan® technology to produce paraffins. These fillers are typically composed of linear paraffins containing 16 to 20 carbon atoms in the form of esters and carboxylic acids, or linked together by a glycerol molecule. Due to their nature, they are ideal raw materials for the production of hydrocarbon fuels in the boiling range of diesel and jet fuel, in addition to a very low sulfur content and a high cetane number.

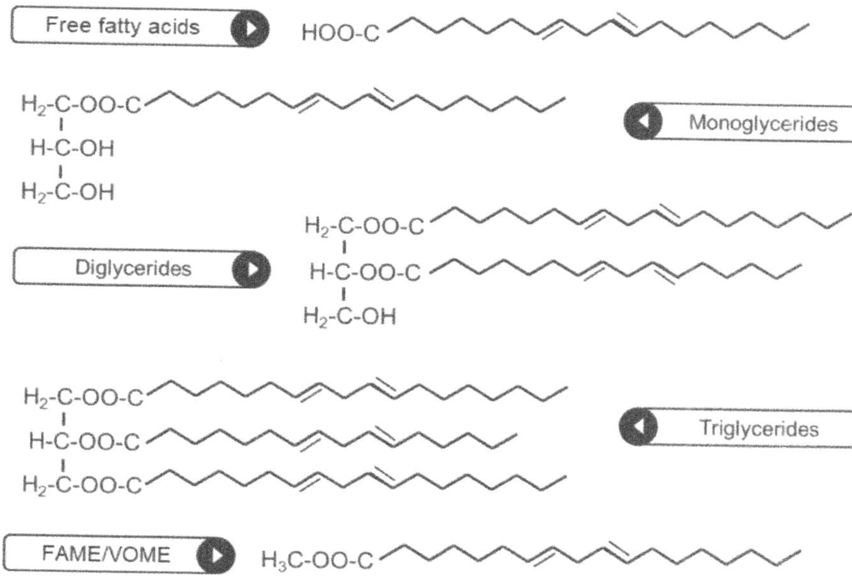

Figure 2.38. *Details of a lipid feed*

The basic structure of these raw materials is summarized in Figure 2.38. They mainly contain lipids, in particular free fatty acids, monoglycerides, diglycerides, and triglycerides. Esters such as fatty acid methyl esters or vegetable oils (FAME/VOME) can also be part of the feed.

Technology

Vegan® technology consists of two sections (Figure 2.39):

1) **hydrotreatment** makes it possible to produce saturated and deoxygenated carbon chains by adding hydrogen. The paraffins thus created have a very good cetane number but limited cold properties;

2) the **hydro-isomerization** step makes it possible to improve the cold properties of the paraffins produced in hydrotreatment, as well as to adjust the targeted cold properties in the product so as to produce biojet fuel and/or renewable diesel.

Figure 2.40 details the evolution of product quality during the different stages of the Vegan® technology.

Through its hydrotreatment step, which allows for the deoxygenation of the lipid molecule, the product resulting from the technology, called renewable diesel or HVO (hydrotreated vegetable oil) is distinguished from biodiesel, which is usually produced *via the transesterification of vegetable oils where the molecule obtained at the end of production always retains an oxygenated part.*

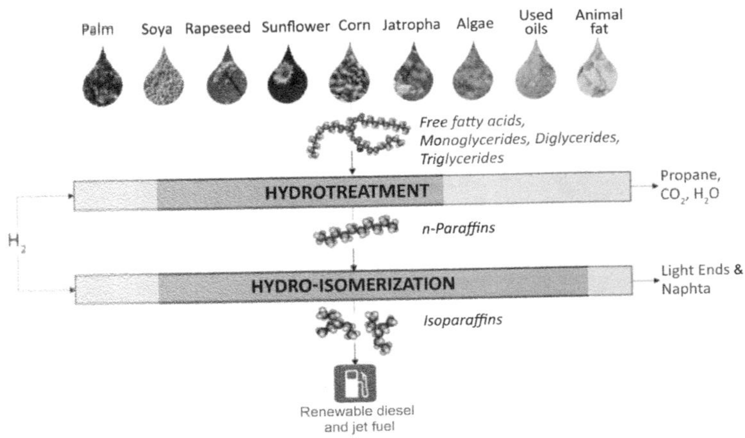

Figure 2.39. *Schematic diagram of Vegan® technology. For a color version of this figure, see www.iste.co.uk/dalpont/process1.zip*

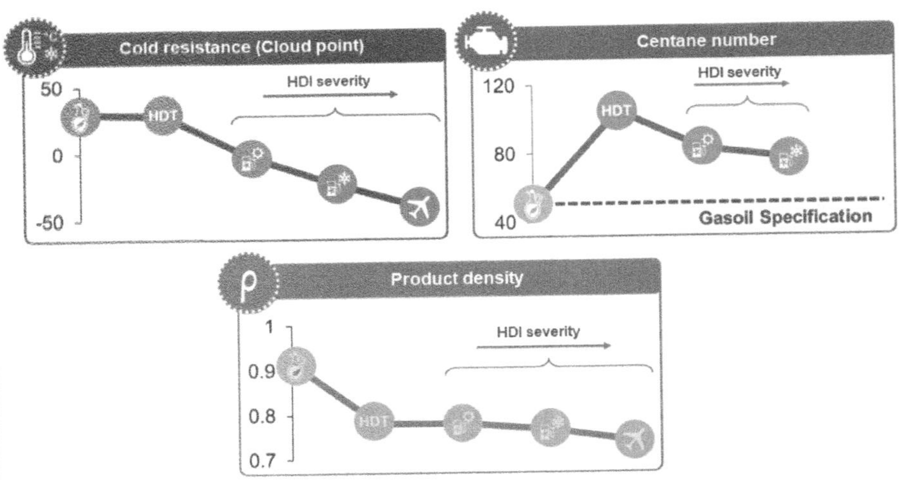

Figure 2.40. *Evolution of product quality during various stages of the Vegan® technology. For a color version of this figure, see www.iste.co.uk/dalpont/process1.zip*

As a result, HVO has characteristics that are clearly superior to that of biodiesel, particularly in terms of cold resistance, cetane number and aging resistance. Another advantage of HVO is its technological flexibility that makes it possible to produce biojet fuel.

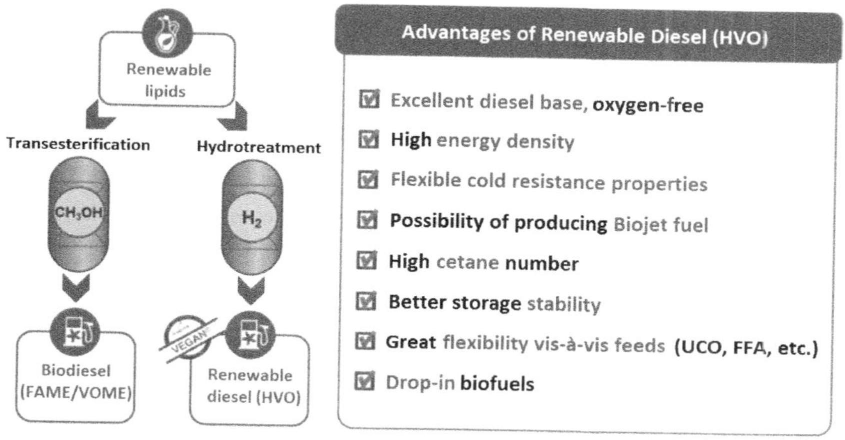

Figure 2.41. *Advantages of HVO product compared to biodiesel (FAME/VOME). For a color version of this figure, see www.iste.co.uk/dalpont/process1.zip*

Description of the Vegan® Process

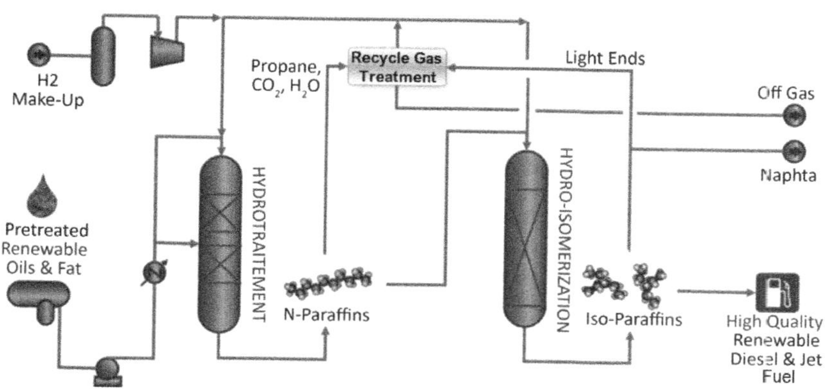

Figure 2.42. *Vegan® process. For a color version of this figure, see www.iste.co.uk/dalpont/process1.zip*

The feed from the unit containing the lipids is pumped and mixed with hydrogen to feed the trickle-bed reactor in hydrotreatment.

The trickle-bed reactor is particularly suitable for this application because, fed into the mixed phase, it acts as a plug flow reactor thus making it possible to obtain a homogeneous reaction gradient along the reactor. Moreover, its configuration is not very complex, making it easy to build which results in moderate costs.

The strong exothermic reaction requires positioning several catalyst beds, separated by quench boxes, to reduce the temperature increase.

To ensure optimal liquid/gas distribution throughout the catalyst bed, the reactor is equipped with EquiFlow™ liquid/gas distributor trays above each of the catalyst beds.

The reactor effluent is cooled, thus making it possible to separate light compounds from the paraffins formed. The gas phase is sent to a processing facility and then partially recycled in the reactors.

The paraffins, free of light compounds, are then sent to the hydro-isomerization reactor, the operating conditions of which are adjusted according to the desired products (diesel or jet fuel).

Like hydrotreatment, hydro-isomerization reaction takes place in the mixed phase. In order to optimize the distribution of the gas and liquid phases, the reactor is also fitted with an EquiFlow™ tray.

The hydro-isomerization effluent is cooled. The gas phase is sent to the same processing facility as the hydrotreatment, before being partially recycled.

The renewable diesel/biojet cut produced is finally stabilized before being sent to storage.

Vegan®: Well-established advanced technology

Vegan® technology is based on the know-how of Axens, in the fields of hydrotreatment and hydro-isomerization/hydrocracking. By adapting its industrially proven technologies and with more than 200 references around the world, Axens knew how to benefit from all of its experience, to optimize and increase the reliability of the industrialization of the process by developing catalysts with high activity and selectivity, by designing an optimized and robust process diagram, as well as developing reactor internals to allow performance to be maximized.

Catalysts

The catalysts used in Vegan® technology are those that have already proven themselves in industrial hydrotreatment units for various applications. These catalysts have been selected and have been the subject of in-depth tests (performance, resistance to impurities, etc.) in the

IFPEN pilots, under representative conditions and on real feed. These catalysts are manufactured and supplied by Axens.

Process

In designing the Vegan® process, process engineers paid particular attention to energy integration in order to minimize the carbon footprint of the process.

To optimize energy integration, the engineers used several methods, including the Pinch method, which consists of analyzing the network of hot and cold fluids and optimizing heat exchanges. By coupling this method with tools developed by Axens, the Vegan® process has an advanced energy integration.

Reactor Internals: EquiFlow™

The hydrotreatment and hydro-isomerization reactors operate in the mixed phase on a trickle-bed.

In order to ensure optimal liquid/gas distribution throughout the catalyst bed, Axens has developed and perfected the EquiFlow™ distributor trays, which equip each catalyst bed with the Vegan® process.

EquiFlow™ distributor trays use a dispersive system located under chimneys to ensure that the liquid/vapor distribution in the catalyst bed is close to ideal.

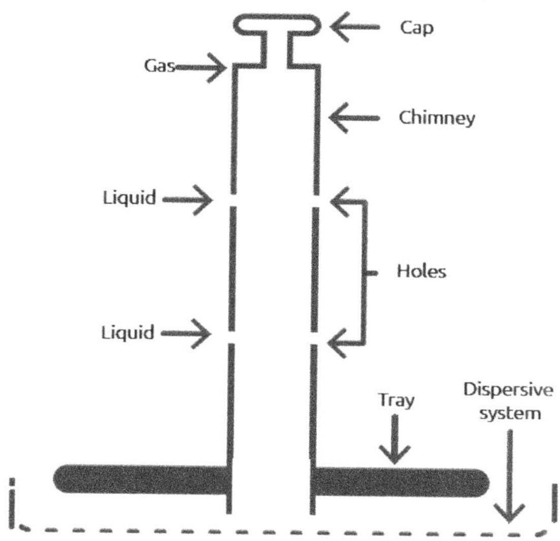

Figure 2.43. *Schematic diagram of an EquiFlow™ Hy-Tray™ dispenser*

Figure 2.44. *3D view of an EquiFlow™ tray installed in a reactor. For a color version of this figure, see www.iste.co.uk/dalpont/process1.zip*

EquiFlow™ distributor trays thus allow better homogenization of the temperature in the catalyst bed, ensuring optimal activity and selectivity of the catalyst, as well as longer cycle times.

Industrial deployment of technology

For the reconversion of its La Mède refinery, Total selected Vegan® technology.

The facility, designed by Axens, is the first of its kind in France and one of the largest in the world, with a capacity of 500,000 tons of renewable diesel per year. The facility was set to operate in 2019 and would supply to the French market.

Find out more about Axens and Vegan®

https://www.axens.net/

Videos

Axens YouTube channel:

https://www.youtube.com/channel/UC9C7x10rC3BEl8P3vjYzD0Q

Presentation of Axens:

https://www.youtube.com/watch?v=UBQvgXW0ckE

Process engineer at Axens:

https://www.youtube.com/watch?v=vF4BMVm89QI

EquiFlowTM:

https://www.youtube.com/watch?v=6H-IY8ZiJUo

Axens

The Axens group (www.axens.net) offers a complete range of solutions including technologies, equipment, furnaces, modular facilities, catalysts, adsorbents, and services for the conversion of petroleum and biomass into clean fuels, as well as for the production and purification of the main petrochemical intermediates. Its services also cover all natural gas processing and conversion options.

The Axens group is ideally placed to cover the entire value chain, from feasibility studies to the start-up and monitoring of the facility, throughout its lifecycle. Its unique range of solutions guarantees optimal performance with a reduced ecological footprint.

Thomas Mallet

Technology Development Manager – Biofuels and Biobased Chemical Components (Axens)

Thomas Mallet joined Axens in 2005 as Technical and Technological Manager of Biodiesel production processes.

In 2008, he joined the Olefins product line in which he was involved in various technologies aimed at recovering light olefins from steam cracking and FCC such as etherification, heterogeneous oligomerization, and Atol (dehydration of ethanol to ethylene).

Since 2014, he has coordinated research and development activities related to the production of biofuel and green chemistry.

Thomas Mallet holds a chemical engineering degree from the *Institut national des sciences appliquées* in Rouen, France.

Larissa Perotta

Technology Team Manager – Biotechnologies: in charge of green fuels, the Olefins and Gas line, and the Process Licensing unit

Larissa joined Axens in 2010 as a process engineer, where she was able to participate in various refining and petrochemical projects. In 2014, she joined the group dedicated to bio-based technologies in order to promote Axens solutions for the production of alternative fuels and chemical intermediates. In this group, of which she is currently the manager, she was able to work on, notably, technologies for the production of cellulosic ethanol and hydrotreatment of lipids.

She holds a degree in chemical engineering from the Federal University of Parana and a master's degree in chemical engineering and refining from the IFP School.

Box 2.6. *Biofuels: a response to energy transition (Thomas Mallet and Larissa Perotta)*

Introduction to product engineering

Product engineering combines formulation and chemical engineering and is defined as the set of operations necessary for the preparation of a product with use value, by mixing and structuring synthetic or natural raw materials.

If we refer to other branches of science such as medicine or astronomy, chemical engineering is a fairly recent discipline, with the first lectures given in 1887 by George E. Davis at the Manchester Technical School in the UK and the first course given in 1888 at MIT by the chemistry professor, Lewis M. Norton. It is also interesting to note that the birth of a discipline is associated with its teaching. After starting from the angle of industrial chemistry, where the manufacturing processes of each specialty were described in detail, two major paradigms – unit operations and coupled transfer modeling – contributed to the success of chemical engineering by unifying the approaches; they are still taught (Favre *et al.* 2008; Hill 2009).

Since the 1990s, the question of the evolution of chemical engineering and its boundaries has been regularly addressed and the concept of product engineering has emerged (Costa *et al.* 2006; Hill 2009). It appears in the Amundson Report (1988), which notes the rapid evolution of the number of products sold on the market. These products are generally consumed, not for their composition, but for their quality and performance, which themselves depend on the composition and structure of the product. It then becomes necessary to develop tools and methods to design and manufacture these new products.

Thirty years later, we can actually see the explosion in the number of products offered to the consumer; very few of these products are "simple" or "raw", that is to say made of a single raw material and having undergone no formulation. In fact, most products are in fact sought after for their useful properties: ease of use, efficiency, multifunctional or sensory aspects. These properties come, of course, from their chemical composition and also from their structure, and it is therefore necessary to question the scientific methodology to be set up to manufacture these useful properties, choose ingredients and manufacturing process(es), and finally characterize these properties.

How to define product engineering

To define the contours of product engineering, it is interesting to start from the evolution of the chemical industry and its needs, particularly in Western countries. According to Amundson (1988), Villadsen (1997), and Wei (2008), the following points can be noted:

– manufactured products evolve rapidly and are frequently replaced by new, more efficient products;

– from homogeneous products with simple molecules, we move towards composite and structured products, manufactured from large molecules with complex structures;

– formulations are reviewed to eliminate unwanted molecules while maintaining performance;

– the commodities industry gives way to an industry of specialty products with high added value, and the industry of mature products and commodities will be relocated to lower-cost countries;

– expertise in product design must be added to that in process design;

– large capacity, single-product continuous units give way to multi-purpose batch plants handling small quantities;

– models become more complex and must be resolved rigorously;

– the need for analytical techniques increases and becomes more sophisticated;

– research evolves towards multidisciplinary teams and studies take into account the micro- and mesoscales, and not only the macro aspects.

This evolution of the commodities industry towards that of high value-added products therefore requires suitable methods and tools in order to tackle these new problems, because it is no longer just a question of producing with a certain purity, maximum yield, at the lowest cost, and in compliance with quality and safety standards, but to guarantee a useful property (Edwards 2006; Hill 2009; Cussler and Moggridge 2011).

Historically, developing new products was rather the exclusive domain of chemists and physicochemists, materials experts, or even the food industry (Hill 2004), and chemical engineering would then take care of process design and development. However, designing multifunctional and multistructured products requires a multiscale approach with consideration of the interactions between various levels (Kind 1999; Wintermantel 1999; Charpentier 2000; Costa *et al.* 2006). Figure 2.45 summarizes all of these scales, taking the elementary production unit, that is to say the equipment, as a central point in the evolution of complexity. We can thus point out that in many fields, the evolution of research tends to go towards the study of microscales, either by the use of digital fluid mechanics and associated visualization techniques, or by the development of microreactors that aim to control the phenomena of transfer and transport at these scales.

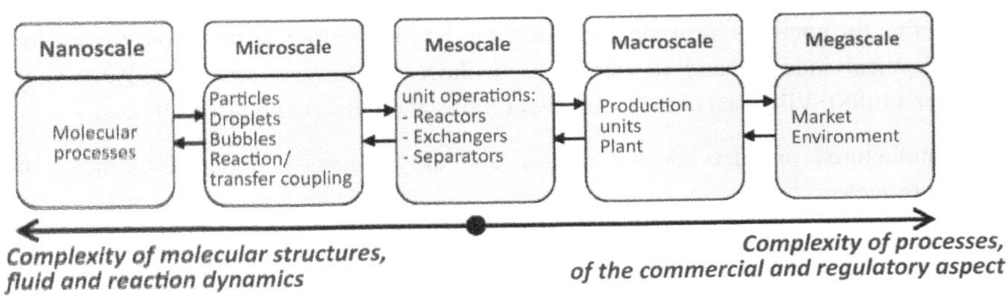

Figure 2.45. *Representation of the scales involved in product design*

Product design is not exempt from this approach since functionality is linked to the nature of molecules, structure to the shape of the entities and interfaces created on a microscale, but manufacturing requires mastery of the equipment used. The other two scale levels are more organizational and strategic than purely technical.

It is therefore a question of developing product-centric chemical engineering instead of chemical engineering centered on the process. A close link with chemistry and physicochemistry is therefore essential and the development of know-how in the area of surfaces and interfaces is necessary to explore structure-property relationships (Amundson 1988). However, process knowledge is essential, because even if the objective here is not the optimization of the process itself, one must be able, with the objective of manufacturing a product, to choose the process or processes that will lead to the desired properties in an optimal way. The approach must therefore be transversal and multidisciplinary (Hill 2009) and, through its history, chemical engineering is already multidisciplinary because it integrates knowledge at the molecular, physicochemical and interfacial levels, more global knowledge of transfer and transport, as well as process control and management. Product engineering therefore appears to be an extension of chemical engineering, not a revolution (Cussler and Wei 2003). However, its evolution can only be done in connection with other disciplines such as organic or inorganic chemistry, science of interfaces, biology, or even mechanics, as already practiced by companies via the concept of project teams (Favre *et al.* 2002).

Useful property concept

Insofar as products are defined not according to their composition, degree of purity, thermophysical, or transport properties, but according to their performance, functionality, and capacity to meet a need, their main characteristic then becomes the useful property (Figure 2.46).

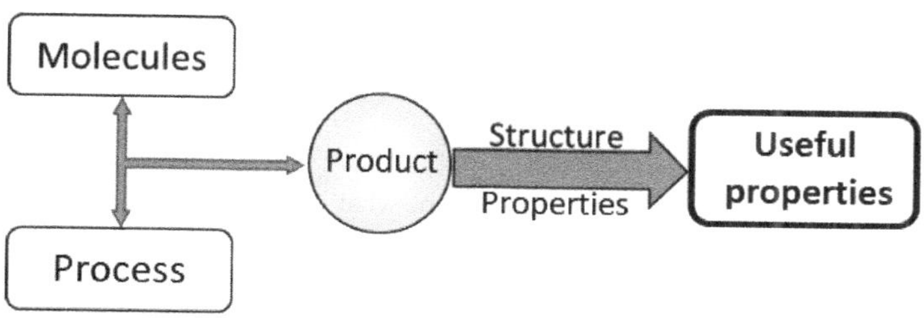

Figure 2.46. *End use properties resulting from formulation and process implementation*

However, to achieve the desired quality, formulations can contain from two to several tens of different compounds, not all being miscible with one another. Interfaces then play a predominant role and give the product its structure. Obtaining this structure very often requires the use of several specific successive unit operations (emulsification, granulation, extrusion, compression, coating, etc.) in place of conventional unit operations, such as distillation or extraction, operations which generally have to be carried out in a precise order to reach the desired final state. The concept of useful property allows us to highlight the rationale involved in the "product engineering" approach, which involves choosing the process and the formula process according to the desired properties, and not in considering the useful properties as the result of the dimensioning of the process. It becomes necessary to develop reverse engineering of product design (Figure 2.47).

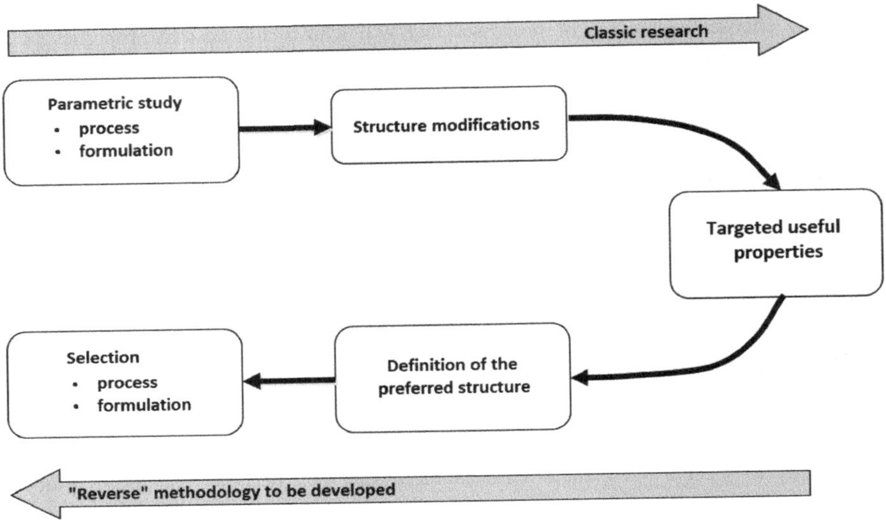

Figure 2.47. *Product development process*

What are the proposed methodologies?

As the problem of product design has arisen, what are the tools to set up in order to achieve the objective in a rational way? This is to avoid the trial and error approach and to provide the most general methodology possible because, at present, the knowledge and associated revenues are part of the proprietary know-how of companies, are often specific to a type of product, and are kept a secret. The development of these design methods, however imperfect they may be, is also an opportunity to highlight the chemical and physical processes not yet described, as well as the needs, particularly in the field of measurement and control of product quality.

Several authors have proposed to break the process down into stages (from 3 to 8 stages): Wibowo and Ng (2001a, 2001b, 2002), Costa *et al.* (2006), Ng *et al.* (2006), Bernado *et al.* (2007), Wesseling *et al.* (2007), Wei (2008), Hill (2009), Cussler and Moggridge (2011). We will choose the proposal of Cussler and Moggridge, who laid the foundations for this approach in 2001, by publishing the first work in the field. The procedure is broken down into four steps, which are summarized and commented on in Table 2.3.

Step	Objectives	Tools	Comments
Needs	Identify client needs, especially those not met	Interviews of clients Interpret clients' needs in technical specifications Comparison to reference products	Difficulty in converting requests in technical specifications
Ideas	Generate a list of products capable of meeting specified needs	Generation of ideas (clients, project team, consultants, competitors) Chemical techniques (screening, combinatorial chemistry) Classification, ordering, and sorting of ideas	General need to convert ideas without constraint
Selection	Choose the most promising idea	Use of thermodynamics Use of kinetics (reaction or transfer kinetics) Use of criteria matrix Risk evaluation (= level of probability x level of impact)	Need for subjectivity Lack of relationship between structures and quantitative properties
Manufacturing	Choose amd design the production process	Intellectual property Research on missing information Development of final specifications Technology choice Scale-up Economic evaluation	Usual difficulty in designing especially with some techniques

Table 2.3. *Steps of the "product engineering" methodology proposed by Cussler and Moggridge (2011)*

The authors mention three main categories of products at stake: commodities, molecular products (also called specialty products), and microstructure products (also known as performance products. They also include the devices performing a chemical modification. The current evolutions and improvement of the approach essentially aim to set up tools that make it possible to rationally detail and address each stage of the design process, starting from the need expressed by the user. Bagajewicz (2016) therefore proposed to take users into account by modeling their preferences. Serna-Rodas (2018) used the Kano model and QFD (Quality Function Deployment) (Ji *et al.* 2014) to analyze and translate requirements into specifications and applied the knowledge base built to development a cosmetic cream. Ng and Gani (2019) made an inventory of existing modeling tools or tools to be set up in order to master the design and production path. Their approach is, however, presented from a top-down angle, that is to say, initialized by knowledge of ingredients and processes, with the dimension of innovation still to be integrated.

Conclusion

Product engineering thus appears to be one of the current developments in chemical engineering. The lack of availability of conceptual tools does not yet allow us to consider that this is a third paradigm, and quantitative description and mastery of usage functions in reverse engineering remain challenges to be met. This objective requires taking an interest in all aspects of a product: chemical composition, thermodynamic and physicochemical properties, interactions with the process, particularly in the creation of microstructures. However, the very wide variety of products involved entails separating them by family. The notion of state of matter (liquid, paste or gels, and solid) proposed by Favre *et al.* (2002) seems relevant because it does not introduce too much granularity into the definition of families, which are thus not limited to an industrial sector and can become the place of synergies linked to similar problems. The research efforts to be carried out are now on three levels:

– quantitatively defining usage functions taking into account users, chemical aspects and processes;

– solving the opposite problem, that is to say identifying necessary ingredients and processes to be implemented to manufacture the desired useful property;

– integrating environmental protection through the rational use of raw materials and energy resources (Woinaroschy 2016).

For more information

Cussler, E.L., Moggridge, G.D. (2011). *Chemical Product Design*. Cambridge University Press, Cambridge.

Hill, M. (2009). Chemical product engineering – The third paradigm. *Computers and Chemical Engineering*, 33, 947–953.

Ji, P., Jin, J., Wang, T., Chen, Y. (2014). Quantification and integration of Kano's model into QFD for optimising product design. *International Journal of Production Research*, 52, 6335–6348.

Ng, K.M., Gani, R. (2019). Chemical product design: Advances in and proposed directions for research and teaching. *Computers and Chemical Engineering*, 126, 147–156.

Woinaroschy, A. (2016). A paradigm-based evolution of chemical engineering. *Chinese Journal of Chemical Engineering*, 24, 553–557.

Véronique Falk

Véronique Falk is a professor of chemical and product engineering at the University of Lorraine. Her research focuses on shaping powders and, notably, on the link between useful properties of products and the process/formulation relationship.

Box 2.7. *Introduction to product engineering (Véronique Falk)*

2.13. References

Amundson, N.R. (1988). *Frontiers in Chemical Engineering: Research, Needs and Opportunities*. National Academy Press, Washington.

Bagajewicz, M. (2016). Integrated consumer preferences and price/demand-driven product design: An alternative to stage-gate Procedures. *Computer Aided Chemical Engineering*, 39, 45–59.

Bernardo, F.P., Costa, R., Saraiva, P.M., Moggridge, G.D. (2007). Chemical product engineering and design: Active learning through the use of case studies. *International Conference on Engineering Education*. Coimbra, Portugal, September 3–7.

Charpentier, J.C. (2000). Did you say: Chemical, process and product-oriented engineering? *Oil & Gas Science and Technology*, 55(4), 457–462.

Costa, R., Moggridge, G.D., Saraiva, P.M. (2006). Chemical product engineering: An emerging paradigm within chemical engineering. *AIChE Journal*, 52, 1976–1986.

Cussler, E.L., Wei, J. (2003). Chemical product engineering. *AIChE Journal*, 49, 1072–1075.

Dal Pont, J.-P. (2012). *Process Engineering and Industrial Management*. ISTE Ltd, London and Wiley, New York.

Dal Pont, J.-P., Azzaro-Pantel, C. (2014). *New Approaches to the Process Industries*. ISTE Ltd, London and Wiley, New York.

Daniellou, F., Naël, M. (1995). *Ergonomie*. Techniques de l'Ingénieur, Saint-Denis.

Edwards, M.F. (2006). Product engineering: Some challenges for chemical engineers. *Chemical Engineering Research and Design*, 84(A4), 255–260.

Favre, E., Marchal-Heussler, L., Kind, M. (2002). Chemical product engineering: Research and educational challenges. *Transactions of the Institution of Chemical Engineers*, 80(A), 65–74.

Harari, Y.N. (2014). *Sapiens: A Brief History of Humankind*. Harvil Secker, London.

Hill, M. (2004). Product and process design for structured products. *AIChE Journal*, 50, 1656–1661.

Kind, M. (1999). Product engineering. *Chemical Engineering and Processing*, 38, 405–410.

Malmströmg, G.B. (ed.) (1997). *Nobel Lectures: Chemistry (1991–1995)*. World Scientific, Singapore.

Martin, S. (2019). *Droit et pratique des emballages. Normes d'écoconception des emballages*. Techniques de l'Ingénieur, Saint-Denis.

Ng, K.M., Gani, R., Dam-Johansen, K. (2006). *Chemical Product Design: Towards a Perspective Through Case Studies, Volume 23*. Elsevier, Amsterdam.

Serna Rodas, J. (2018). Methodological approach for the sustainable design of structured chemical products during early design stages. PhD thesis, University of Lorraine.

Villadsen, J. (1997). Putting structure into chemical engineering. *Chemical Engineering Science*, 52(17), 2857–2864.

Wei, J. (2008). *Product Engineering: Molecular Structure and Properties*. Oxford University Press, Oxford.

Wesselingh, J.A., Kiil, S., Vigild, M.E. (2007). *Design and Development of Biological, Food and Pharmaceutical Products*. John Wiley & Sons, Chichester.

Wibowo, C., Ng, K.M. (2001a). Product-oriented process synthesis and development: Creams and pastes. *AIChE Journal*, 47, 2746–2767.

Wibowo, C., Ng, K.M. (2001b). Operational issues in solids processing plants: Systems view. *AIChE Journal*, 47, 107–125.

Wibowo, C., Ng, K.M. (2002). Product-centered processing: Manufacture of chemical-based consumer products. *AIChE Journal*, 48, 1212–1230.

Wintermantel, K. (1999). Process and product engineering: Achievements, present and future challenges. *Chemical Engineering Science*, 54, 1601–1620.

3

Designing Chemical Products

3.1. Introduction

3.1.1. *Why is chemical product design important?*

Following the definition by Kotler, a product is something which can be offered to a market in order to fulfill the requirements of a customer (Kotler 2006). This very fundamental and broad definition of a product already includes the most important issue: the customer's needs and requirements. In the middle of the last century, the markets were created and driven by the new technologies and developments that came merely from science and engineering, such as new materials like polymers which were developed without any preparatory marketing investigations. New technology leads to a new product which creates a completely new market and also new employment. People often call this the technology push area.

Looking at the markets coming out of the chemical engineering field today we have to observe that markets are well established and quite saturated. Bulk chemicals and commodity businesses are characterized by rationalization of workflows to reduce costs to be competitive. The recently increasing prices of raw materials will make this business even harder.

Looking at specialty chemicals we can also see trends of a cut-throat competition of production technologies. It is likely that chemical product routes get more and more substituted by bio-based processes which are getting more and more competitive. The production of riboflavin, vitamin B2, is such an example. The increasing oil prices and the discussion about CO_2 emissions related to global warming issues will strengthen this trend towards a bio-based society. But new

---

Chapter written by Willi MEIER.

production processes in a cut-throat competition market will not lead to an increase of employment. Jobs which are created in the new production processes will be lost by shutting down the old ones.

Many traditional chemical companies have already responded to these new developments and are currently adapting their structure to the new requirements. Over its 100-year history DSM, for example, has transformed from a local coal mining operation, at first, to a chemical commodity producer. After the divestments of the petrochemical activities in 2002 and Bakery Ingredients in 2005, the recent inclusions of DSM Nutritional Products (formerly Roche Vitamins and Fine Chemicals) in 2003 and the recent acquisition of NeoResins, DSM has now a portfolio which comprises around 80% specialties (Laane 2006).

The development within DuPont is similar and is sketched in Figure 3.1. Starting with explosives in the 19th Century, turning to chemicals in the 20th Century, DuPont sees the future in multi-disciplinary knowledge intensive solutions, required by the customers of the 21st Century.

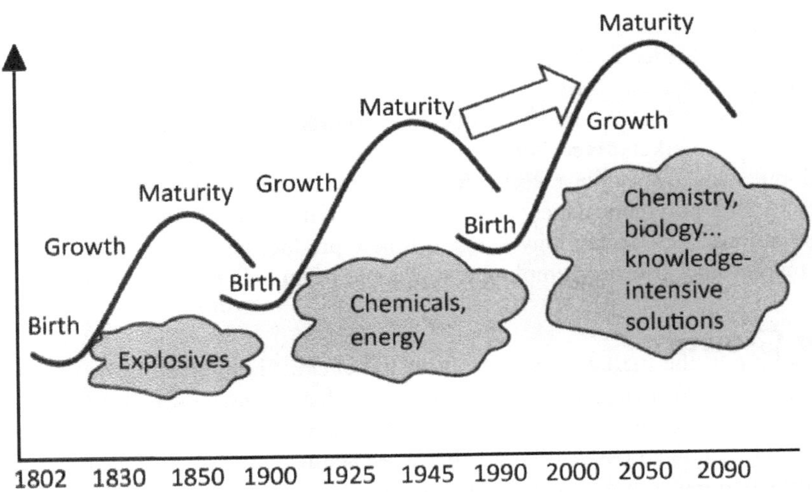

Figure 3.1. *DuPont business lifecycle over time (source: (Connelly 2006))*

DuPont's future strategy clearly shows the importance of Product Design and Engineering in the future of the Process Industries. Customers' demands have to be recognized and turned into products by the help of well-established processes and technologies. Fulfilling customers' needs will automatically lead the business to new markets were real growth and new employment is created. Some people call these blue ocean markets where new fields are opened and competition does not exist (Kim 2006).

3.1.2. *Current state of the art*

There are some general trends which may be observed in all areas of chemical and pharmaceutical industry:

– decreasing number of basic inventions of new molecular structures;

– increasing importance of customer convenience;

– increasing importance of environmental issues and the circular economy;

– shortening of product lifecycles due to increasing competition;

– Big Data, digitalization in research, development and sales.

In the last century product design was driven by the technology push. Examples are the first polymer products, or paints and adhesives in the fifties and sixties. One of the last products in this respect was the invention of the Post-it several years ago, where the company 3M has developed an adhesive which did not perform very well – the binding power was very weak. With the idea of Post-it notes, a new product line was created, which was based on special qualities of the developed adhesive, which was obtained unintentionally.

Nowadays the customer's demands are clearly in the focus of product design.

As a consequence, there is a shift in the way products have to define their unique selling point. In the past it was predominantly a new molecular entity which defined a new product. Nowadays it is a comprehensively designed product which delivers the answer to a customer need. Once this is addressed, the above trends should be understood as a chance, rather than a threat, for the pharmaceutical and chemical industry, since it opens up new ways to distinguish our products. It also implies a structural change from manufacturer to solution provider.

Due to the megatrend of globalization, competition in the chemical industry has raised dramatically within the last few years. Former emerging countries like China or India now play a significant role in the global market. Bulk chemicals as well as specialty chemicals produced in China or India are available at a very competitive price based on the well-known reasons for low cost production in these countries.

Beside the megatrend of globalization, digitalization technologies and the Internet of Things, the development of chemical products is also changing, as are the requirements these products will have to fulfil in the future.

3.2. Basic technologies

3.2.1. *Dimensions*

Product design takes place on different levels of scale. In Figure 3.2 the different levels are shown.

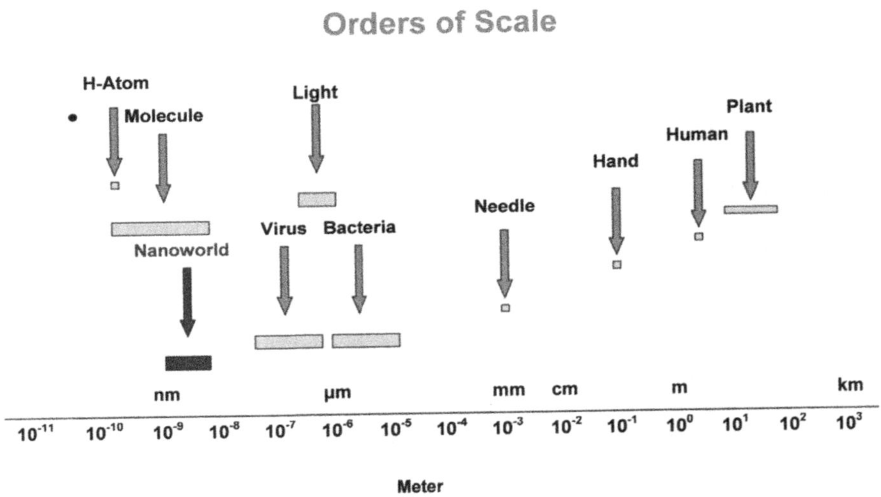

Figure 3.2. *Scales where chemical product design takes place. For a color version of this figure, see www.iste.co.uk/dalpont/process1.zip*

Starting at the atomic and molecular level, which is in the range of nanometers, one reaches the micro levels, where particle technologies like agglomeration are applied to set up the wanted product specifications. At the micro level biotechnology processes also take place. We then reach the macroscopic scale, where products directly interact with the customers. Typical products here are tablets, powders, or even automotive cars.

At all these different scales product properties have to be adjusted to deliver the wanted specifications.

At the molecular level the right active ingredients have to be chosen. This is the playground for chemists who optimize the structure of potential drugs and ingredients to adjust certain product qualities. An example of this is shown in Figure 3.3. The figure shows the molecule acetylsalicylic acid, better known by its common name Aspirin®. Aspirin® is a good example for the development of product design and how this process has changed during the decades. The consumer sees the product, for example, as tablets.

In section 3.3.1 the product design of Aspirin® will be described in more detail.

Figure 3.3 also shows a brown powder and its agglomerated form. This powder is plain coffee and chemical engineers simply adjust solubility parameters of this product by agglomeration.

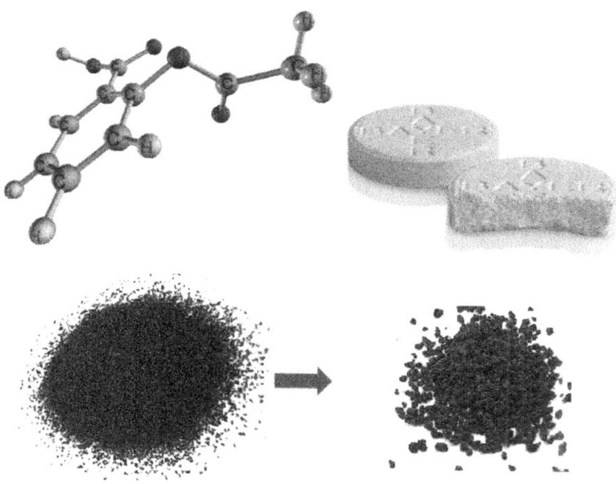

Figure 3.3. *Chemical products in raw form and as customer product. For a color version of this figure, see www.iste.co.uk/dalpont/process1.zip*

3.2.2. *Additives*

Additives are widely used in product design to adjust certain product qualities like solubility or texture. In this section two widely used additives are introduced.

3.2.2.1. *Starch*

Starch is widely used as an active ingredient in product design. It is cheap and starch is the most abundant biopolymer worldwide. In Figure 3.4 the yearly production of starch within the EU is shown.

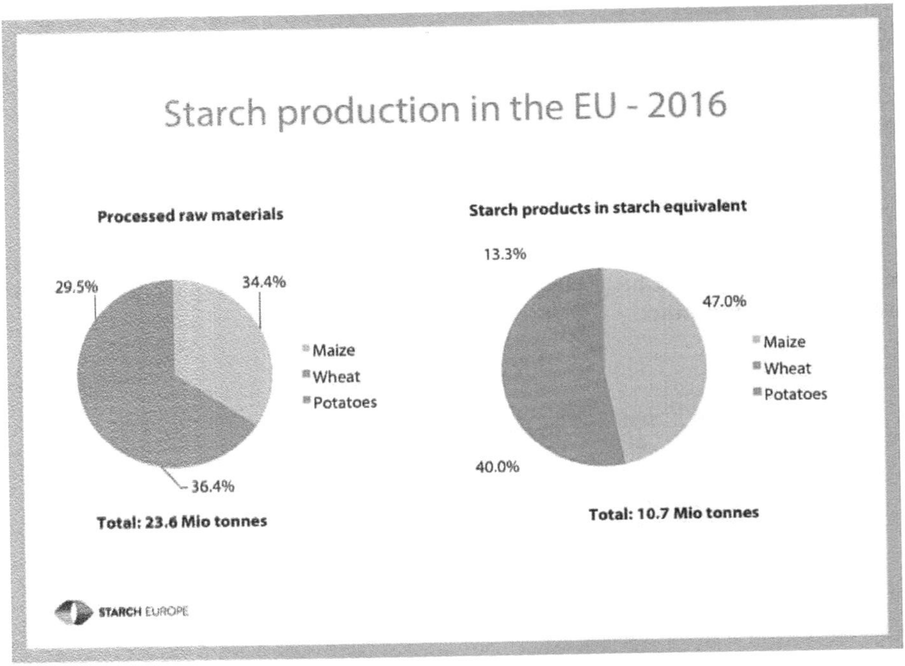

Figure 3.4. *Starch production in the EU in 2016. For a color version of this figure, see www.iste.co.uk/dalpont/process1.zip*

As a water binding and thickening ingredient starch is used to build up texture, improve consistency, increase viscosity and prolong shelf life. It is also found as a filling material, encapsulating or emulsifying agent or even as a fat replacer. Natural or chemically modified starches are widely found in diverse applications in food, packaging, pharmaceuticals and the textile industry (Punia 2019). Figure 3.5 shows the main starch applications.

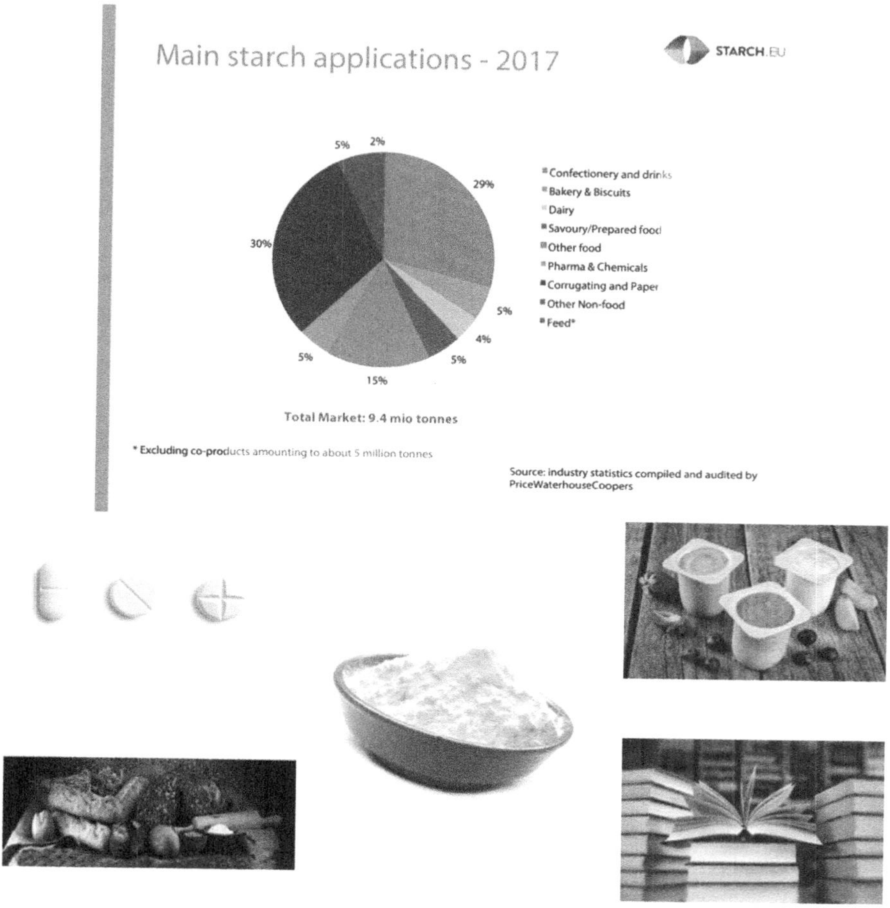

Figure 3.5. *Main starch application in 2017. For a color version of this figure, see www.iste.co.uk/dalpont/process1.zip*

Starch has a unique quality when it is heated. The viscosity increases up to a temperature of 95°C. Then the viscosity goes down. When the starch is cooled down below 95°C viscosity increases again. This behavior is based on the formation of amylose networks, which lead to a thickening of the product. Figure 3.6 shows an amylogramm which demonstrates this behavior. Starch is heated with an increasing temperature of 1.5°C and the viscosity is measured. After swelling the amylose network is formed which lead to a high viscosity consistency of the product. When passing 95°C the viscosity goes down, and after cooling down the amylose network is formed once again.

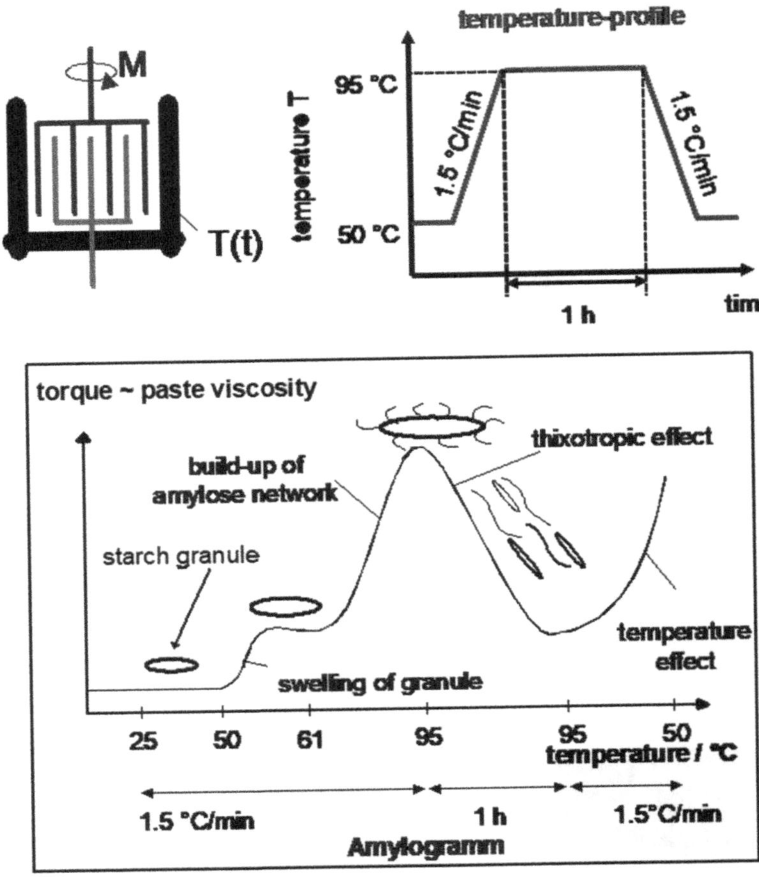

Figure 3.6. *Temperature dependency of starch gelatinization. For a color version of this figure, see www.iste.co.uk/dalpont/process1.zip*

3.2.2.2. *Gelatin*

Gelatin is a derivate of collagen, a principal structural and connective protein of animal origin – it is a natural product consisting of a mixture of polypeptides varying in length ranging from a few thousand Dalton (atomic mass unit 1D = 1.66 10^{-27} kg) up to a couple of hundred thousand D.

In general gelatin is derived from collagenous animal tissues like bovine hide, pig skins and bones, a byproduct of animals subject to government-certified pre- and post-mortem inspections as passed as fit for human consumption by means of acidic, alkaline and thermal treatment of the primary raw materials.

The term “gelatin” has been used since about 1700 CE and is derived from Latin where “gelatus” means “frozen”; related to one of the major applications of gelatin: its ability to form thermo-reversible gels, which is unique for proteins. There is evidence that gelatin has been used for at least 4,000 years by cooking bones or skins; the resulting product was applied due to its adhesive properties for technical applications. Collagen, the starting material for the production of gelatin, is the most abundant component of the extra cellular matrix in mammalian tissues; it accounts for about 30% of its protein content.

Gelatin and gelatin hydrolysates are insoluble in less polar organic solvents such as benzene, hydrocarbons, aliphatic and aromatic alcohols, ketones and ethers. They are easily dissolvable in some polyalcohols like glycerin or sorbitol. Solutions of gelatin in mixed solvent types, like low molecular weight alcohols and ketones and water can be realized by gradually adding, under agitation, the organic solvent to a hot aqueous gelatin solution (up to 50% organic solvent).

One of the most important properties of gelatin is its ability to form heat-reversible gels in water and polyhydric alcohols over a wide range of gelatin concentration, pH values and in the presence of sugars and electrolytes. As an example, if a hot gelatin solution of greater than 0.5% gelatin concentration is cooled to about 20–30°C, the viscosity of the solution initially increases drastically; on further cooling a gel is formed. The rigidity of the gel depends on the temperature gradient during gel formation, concentration and the quality of the gelatin and components that interfere with the gel formation. As an example, sorbitol, fructose or Na_2SO_4 stabilize the helix formation, that is, increase the rigidity of the gel, whereas urea and NaCl have a negative impact on the gel strength.

Gelatin is widely used in the production of food. Different properties and various applications require the selection of the appropriate gelatin type.

For the confectionary industry, gelatin is used due to a multitude of properties:

– for the manufacture of gummy candy and fruit chews, gelatin is responsible for gel formation and texture;

– aerated products like marshmallows contain gelatine to build up and to stabilize foams;

– film forming and binding properties are used for the production of dragees and compressed products;

– in the dairy industry gelatine avoids syneresis; as an example: the separation of whey from yoghurts.

Gelatin is the gelling agent in the production of aspic meat and sausage products. Depending on the gelatin type, the texture of the gel can range from soft to sliceable. In addition, the fat content of canned meat and sausage can be reduced, the spreading properties and the softness of emulsified products can be improved. In beverages, gelatin can assist to clarify wine and fruit juices; in some countries it is used to clarify beer. In bakeries it is often used as a binder for fillings, icings and toppings and for the stabilization of creams and whipped cream.

Among the pharmaceutical uses, gelatin capsules are the best known; it serves as a bulking agent for the production of free-flowing water insoluble vitamins, as a starting material for the manufacture of plasma substitute products and for the production of surgical sponges; it acts as a binder in wet granulation for tablet production. Gelatin is recommended as a food supplement in the case of joint diseases.

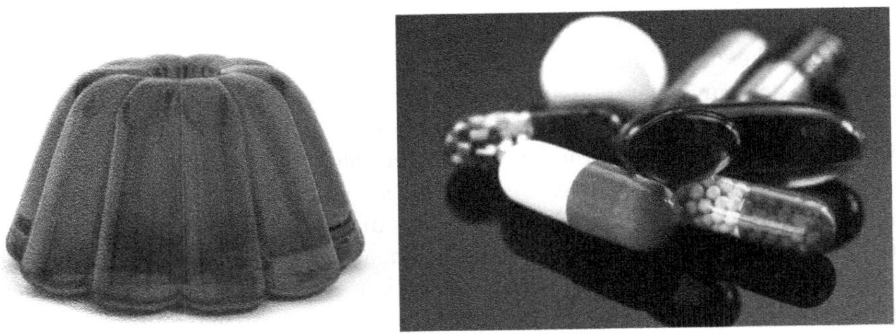

Figure 3.7. *Examples of the application of gelatin in the food and pharmaceutical industry. For a color version of this figure, see www.iste.co.uk/dalpont/process1.zip*

3.2.3. *Microencapsulation*

Encapsulation or coating has been used for many years in many production areas, for example, in the food or detergents industries. Air sensitive ingredients like flavors or fragrances are protected by coatings made of fats or oils. Spray drying of moving bed reactors are the used technologies for such encapsulation procedures.

Microencapsulation is a process by which very tiny droplets or particles of liquid or solid material are surrounded or coated with a continuous film of polymeric material.

Microencapsulation can be defined as the process of surrounding one substance with another substance on a very small scale, yielding capsules from several microns to several hundred microns in size.

Microencapsulation offers a whole new range of product properties. Different capsulation materials can be used to adjust the time of release of the encapsulated material. In the pharmaceutical industry, for example, it is possible to maximize the therapeutic efficacy when the drug is delivered to the right target in the optimal amount and in the right time period. Even highly active drugs are useless when they don't pass the acid environment of the stomach.

One of the most used encapsulation materials are hydrocolloids. Hydrocolloids are water soluble macromolecules of high molecular weight which, by binding a large quantity of water, modify the rheology of aqueous systems to which they are added. Hydrocolloids have the ability to thicken, stabilize or gel aqueous systems

Alginat belongs to this group of substances and is widely used in the food and pharmaceutical industries (Shaharuddin and Muhamad 2015). Alginic acid is produced from black seaweed and its sodium salt is mainly used for encapsulation. Industrial production of alginates is based on two properties: sodium and potassium alginates are soluble in water and alginic acid and its calcium salts have very low water solubility.

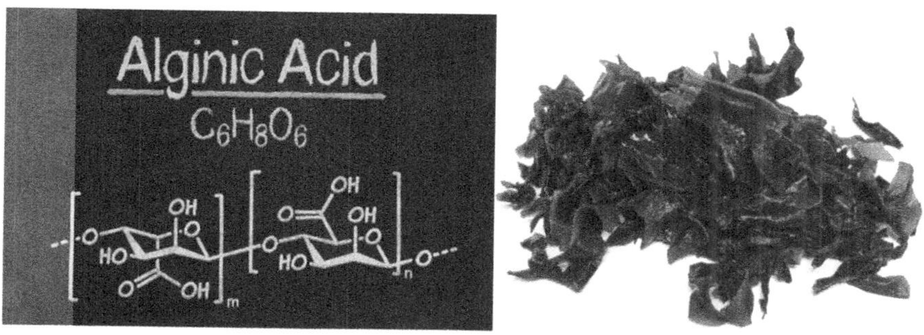

Figure 3.8. *Structure of alginic acid. For a color version of this figure, see www.iste.co.uk/dalpont/process1.zip*

In calcium-free media alginate is completely soluble in water. The viscosity depends on the dosage, the lengths of the chains and the temperature (Figure 3.9a). In media with calcium, gels are formed which are strong, coherent and thermo-resistant.

Adding a sodium alginate solution to a CaCl2 solution which also contains the ingredient which should be encapsulated forms alginate beads directly around the ingredient (Figure 3.10). This process is very easy to apply so it is widely used to produce products in the food, detergents or pharmaceutical areas.

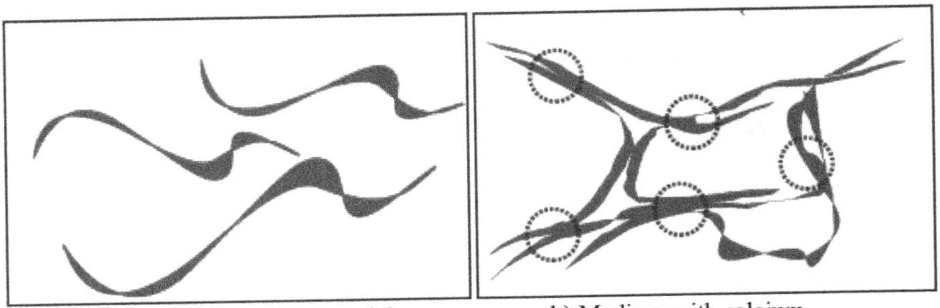

a) Medium without calcium b) Medium with calcium

Figure 3.9. *Structure of alginate solution with or without calcium*

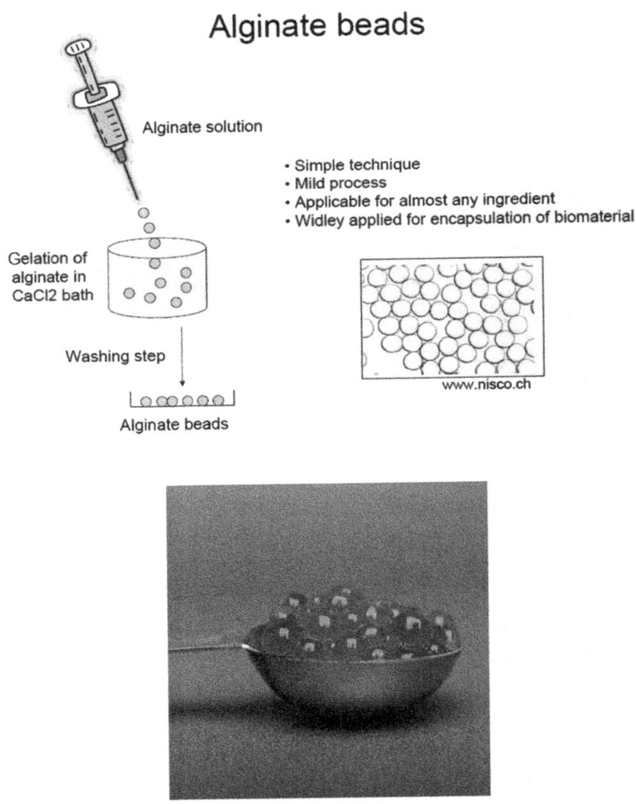

Figure 3.10. *Principle production of alginate beads. For a color version of this figure, see www.iste.co.uk/dalpont/process1.zip*

3.3. Products

3.3.1. *Aspirin®*

Aspirin® is one of the few brands to have maintained a leading market position despite generic competition and is still the object of current research (Chen 2019). The manufacturer Bayer is not resting on its laurels, however, but has further developed the classic. Changes in the composition and galenics of the tablet lead to a much faster onset of action.

The history of Aspirin® is shown in Table 3.1

August 10, 1897	First chemically pure and stable acetylsalicylic acid synthesized by Bayer researcher Dr. Felix Hoffmann
November 1899	Aspirin® sold in 250 g bottles as powder to pharmacies. Pharmacies weigh out 1 g doses and dispense it as sachets to patients
February 27, 1900	US patent granted for the manufacturing process of acetylsalicylic acid
1906	Aspirin® registered as international trademark
1912	Bayer launches first Aspirin® "soluble" tablet with calcium
June 23, 1971	Prof. John Vane (UK) discovers the mode of action of ASA and publishes "Inhibition of Prostaglandin Synthesis as a Mechanism of Action of Aspirin-like Drugs" in the journal *Nature*
1971	Aspirin® plus C effervescent tablet launched as line extension in Germany
1988	Physicians Health study confirms cardio-protective effect of regular ASA intake
1992	Bayer launches Aspirin® Direkt Chewable Tablet as line extension in Germany
1993	Bayer launches Aspirin® Protect 100 mg and 300 mg enteric coated tablets in Germany
2000	Bayer launches Aspirin® Migräne for migraine headaches in Germany
2003	Bayer launches Aspirin® Effect granules in Germany
2003/2004	Bayer launches Aspirin® Complex (ASA plus Pseudoephedrine) in Germany for symptomatic treatment of the common cold
2014	Aspirin® Express, microactive technology. New formulation works two times faster

Table 3.1. *History of Aspirin®*

At the beginning Aspirin® was a simple product consisting of acetylsalicylic acid (ASA). The product design was simple, with starch powder as a binder and disintegrant.

Over the years the base formulation of Aspirin® remained largely unchanged and was gradually improved to accommodate faster tablet presses and new analytical requirements, such as dissolution testing. Significant improvements in compression behavior, flow and disintegration were achieved by partially replacing starch with cellulose derivatives like microcrystalline cellulose or granulated cellulose.

The selected ingredients also support the chemical stability of ASA. This is crucial, since ASA is a very reactive and labile ingredient and the pharmacopoeias allow only 0.3% (w/w) of free salicylic acid for the plain tablet at the end of its shelf-life.

One major improvement was the development of the effervescent tablet. Aspirin® Plus C Effervescent Tablets were developed as a fast acting, convenient dosage form. The active ingredient is already dissolved when it reaches the stomach. Absorption of the drug and, consequently, onset of action are accelerated compared to regular tablets. Furthermore, the free acid is transformed into the gentler sodium salt of acetylsalicylic acid through the effervescent reaction.

Figure 3.11. *100 years of Aspirin® product design. For a color version of this figure, see www.iste.co.uk/dalpont/process1.zip*

One major challenge for the product design concerning Aspirin® has been the improvement of the stomach compatibility. Within our stomach there is a strong acid climate. The hydrochloric acid generates a pH of around 2, which is good for the elimination of unwanted bacteria in our food. To ensure stable pH conditions when taking Aspirin®, a buffer system is added such that stomach compatibility is increased and the CO_2 for the effervescence is produced. Figure 3.12 shows the relevant chemical reactions.

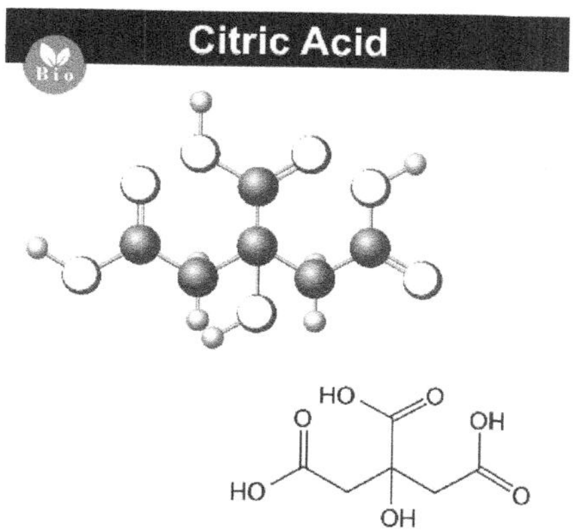

$$\text{Acetylsalicylic Acid} + NaHCO_3 \rightarrow$$
$$\text{Sodium Acetylsalicylate} + CO_2$$
$$\text{Citric Acid} + NaHCO_3 \rightarrow \text{Sodium Citrate} + CO_2$$

Figure 3.12. *Buffer system based on citric acid*

This buffer system with citric acid is widely used in the formulation world, not only in pharmaceutical applications but there are also a lot of applications in the food industry.

In the next chapter one of these applications are described in more detail.

3.3.2. *Coffee and related beverages*

In trading value coffee is second only to crude oil. Green coffee production varies in quantity as all agricultural products do, in the case of coffee, mainly depending on climate conditions in Brazil, the world's largest producer. Contrary to emerging markets for soluble coffee like Russia and China, the consumption of pure soluble coffee in traditional markets is stagnant, while consumption of coffee-based beverages is rapidly growing (Figure 3.13).

Figure 3.13. *Coffee-based beverages consumption 2018 in the US (source: Hamilton Beach). For a color version of this figure, see www.iste.co.uk/dalpont/process1.zip*

The consumption of coffee among the different countries varies across a wide range. Figure 3.14 shows the distribution of coffee consumption among some selected countries.

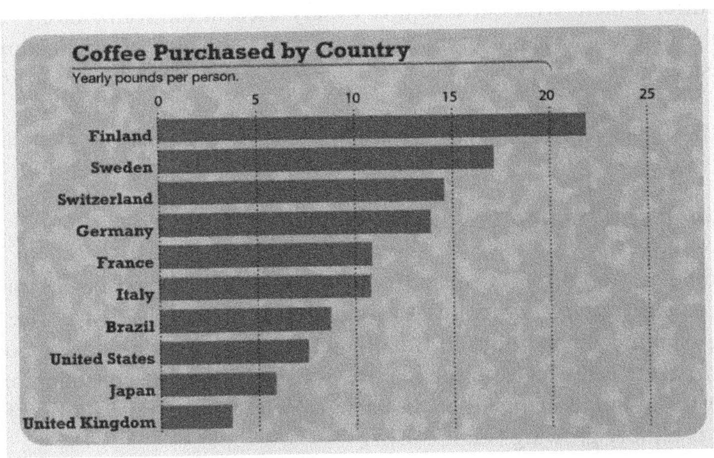

Figure 3.14. *Distribution of coffee consumption in 2018 (source: Hamilton Beach)*

Coffee grows best in a warm, humid climate with a relatively stable temperature of about 27°C all year round. The world's coffee plantations are thus found in a wide, horizontal band on both sides of the equator, between the tropics of Cancer and Capricorn, known as "the coffee belt".

The whole process of creating a powdery coffee-based beverage can be divided into three parts: preparation and extraction of raw materials, creation of a new solid structure (texturization) and, finally, mixing it with non-coffee materials. The red coffee cherries are harvested, shipped and then get roasted, where the aromatic compounds are formed. After milling, spray drying or freeze drying methods are applied to generate the right agglomeration level to adjust the wanted solubility parameters of the product.

The third step of the product is nowadays the most interesting one, because, as mentioned before, the consumption of coffee-based beverages like cappuccinos, lattes, macchiatos etc. is heavily increasing (Ciaramelli 2019).

Figure 3.15. *Coffee: from raw material to new design products. For a color version of this figure, see www.iste.co.uk/dalpont/process1.zip*

One problem that occurs during the formulation of these new coffee products is the acidity of the natural coffee. Many of the products contain milk powders which will lead to flocculation after the dissolution. The pH of a coffee solution lies in the range of 4.7–5.2. Flocculation cannot be tolerated because customers won't accept such unattractive products.

The solution to this is the implementation of a buffer system similar to the one used for the Aspirin® Plus C Effervescent Tablet. Citric acid is used to stabilize the pH environment such that milk and other dairy products do not flocculate when dissolved with coffee powder in water.

Figure 3.16. *Cappuccino with and without buffer system*

3.4. Product design 4.0

We are living in interesting and exciting times. As mentioned in section 3.1.2 there are megatrends in technology and society which have an impact, beside all other issues, on the way we design our products. The normal product lifecycle is shown in Figure 3.17.

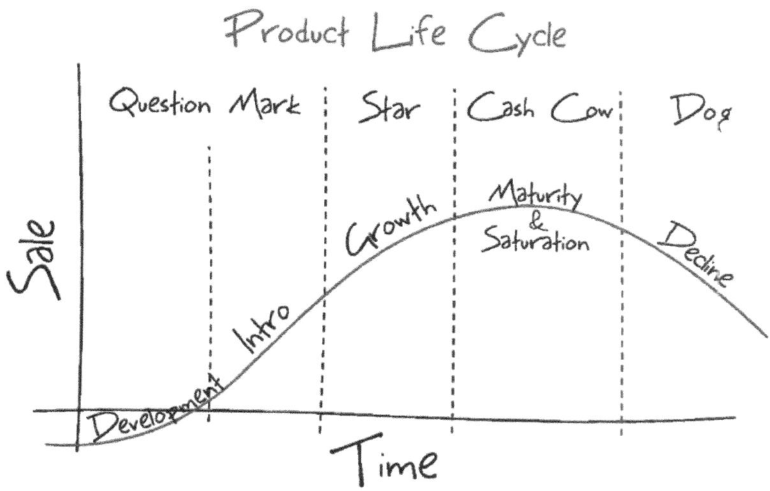

Figure 3.17. *Product lifecycle*

After the development phase, where investments have to be made, the introductory period starts. Here early adapters are the main buyers for this product. Think of smartphones in the early nineties. Only a few people saw the advantage of these products, while the great majority neglected this trend. The next phase is called the growth phase, where nearly everybody wants to have this product. In the case of smartphones this was the case during the last decade. The product now belongs to the "star group" where market share and turnover is increasing. Seeing the latest sales figures for Apple's iPhone one can now say that the product now reaches the maturity & saturation period. Everybody who can afford the product owns one or more smartphones. The next stage will be the decline. This could be started when smartphones are substituted by other devices like intelligent glasses or watches.

Another buzzword which is used in this context is disruptive. Take here as an example the development from vinyl records over music tapes to CDs and DVDs, USB sticks and finally cloud services, where music is not a product but a service.

As chemical products like polycarbonate are not end user products one might not see the consequences of such a development for the related industry. However, if you know that CDs or DVDs are made out of polycarbonate, you recognize the consequences.

Only the last step can be called disruptive because up to that step the business stays the same, only other products were sold in the shops. However, with the change from product to service the whole distribution system collapsed. You see it clearly when you remember the stores of Virgin records where mainly vinyl records and CDs were sold. Nowadays it is more a place to network, with food and party areas.

For bulk chemicals and intermediates the distribution channels will be changed dramatically in the future. Online platforms powered by Artificial Intelligence find the cheapest and most reliable suppliers, no need to ask for several quotes. It is likely that this process of purchasing and selling will be done by machines without any human interaction.

Sensors of all kinds became incredibly cheap during the last decade. Packaging with intelligent sensors allow extensive product tracking such that information is available so that the circular economy becomes reality (Moraga 2019). The idea behind the circular economy is that all products must be recycled and used as raw material for new products. Yesterday, product lifecycles comprised of products from the cradle to the grave and today it is from cradle to cradle (Braungart 2014).

Currently researchers in the chemical industry are intensively involved in ways to reduce the use of raw materials and resources and to reduce the gas emissions which are responsible for climate change. Carbon dioxide is the main focus here but gases resulting from agriculture like methane of nitrogen oxides, which are produced by the degradation of fertilizers, are many times more harmful.

Remember, in section 3.2.3, we introduced the technology of microencapsulation. Researchers are working on an encapsulated product which contains the grain, the right amount of fertilizer and the right amount of pesticides. This would avoid the unnecessarily high amount of nitrogen-based fertilizers which are distributed on the fields. The use of pesticides can also be reduced to a minimum.

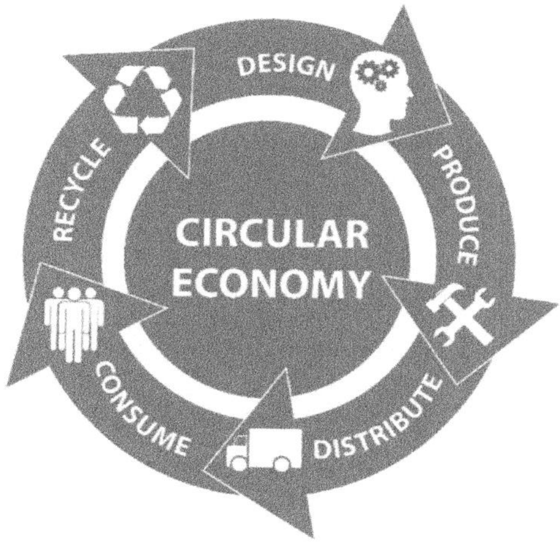

Figure 3.18. *Principle of circular economy*

Other developments are clothes which no longer get dirty. Tailor-made coatings protect the fabrics form dirt etc. by their nanostructure, where liquid droplets or particles have no chance to stick.

The chemical industry, with its tradition of over 200 years, is mature in many areas like bulk chemicals etc. Increasing efficiency is the dominant managing challenge here.

Designing new and intelligent products is still the area where efficiency is the lead managing direction and product design plays a more and more important role in value creation for the industry, environmental protection, including climate change, and well-being for society.

3.5. References

Braungart, M., McDonough, W. (2014). *Cradle To Cradle*. Piper Verlag, Munich.

Chen, Y., Kang, S., Yu, J., Wang, Y. (2019). Tough robust dual responsive nanocomposite hydrogel as controlled drug delivery carrier of asprin. *Journal of the Mechanical Behavior of Biomedical Materials*, 92, 179–187.

Ciaramelli, C., Palmioli, A., Airoldi, C. (2019). Coffee variety, origin and extraction procedure: Implications for coffee beneficial effects on human health. *Food Chemistry*, 278, 47–55.

Kim, W.C., Mauborgner. (2005). *Der Blaue Ozean als Strategie*. Hanser Verlag, Munich.

Kotler, P., Bliemel, F. (2006). *Marketing Management*, 10th edition. Pearson Studium. Munich.

Moraga, G., Huysveld, S., Mathieux, F., Blengini, G.A. (2019). Circular economy indicators: What do they measure? *Resources, Conservation and Recycling*, 146, 452–461.

Laane, C., Sijbesma, F. (2006). Industrial biotech at DSM: From concept to customer. In *Value Creation: Strategies for the Chemical Industry*, Budde, F., Felcht, U.-H., Frankemölle, H. (eds). Wiley VCH, Weinheim.

Punia, S., Siroha, A.K., Sandhu, K.S., Kaur, M. (2019). Rheological behavior of wheat starch and barley resistant starch (type IV) blends and their starch noodles making potential. *International Journal of Biological Macromolecules*, 130, 595–604.

Shaharuddin, S., Muhamad, I.I. (2015). Microencapsulation of alginate-immobilized bagasse with Lactobacillus rhamnosus NRRL 442: Enhancement of survivability and thermotolerance. *Carbohydrate Polymers*, 119, 173–181.

4

Chemical Engineering: Introduction and Fundamentals

4.1. Introduction: definitions, history, and challenges

Chemical engineering is the science of transforming materials and energy on an industrial scale: it is thus the body of knowledge and know-how necessary for designing, implementing, and optimizing industrial product manufacturing and energy production processes. These processes incorporate the equipment in which the physico-chemical and biological transformation of raw materials into functional products (with useful properties) and energy is carried out on an industrial scale.

Managing resources; protecting the environment; controlling safety; and safely, efficiently, and economically producing sustainable products that are useful every day: these are the missions of chemical engineering.

All engineers use mathematics and physics to solve technical problems, but only chemical engineers also use chemical and biological sciences to design and improve the industrial processes that supply the products and energy used in our daily lives: chocolate, gasoline, drugs, plastics, drinking water, electricity, etc. The list is endless. Chemical engineering also deals with recycling effluents, reducing discharges and waste, and reducing industrial energy consumption.

Chemical engineering careers offer excellent job prospects, some of the highest salaries in scientific careers, a diversified job made up of innovation and challenges, openness to the world and to other activities, and the opportunity to make a substantial contribution to a sustainable future. Thanks to the skills acquired during their training, technicians, engineers, and researchers in chemical engineering are

---

Chapter written by Marie DEBACQ, Alain GAUNAND and Céline HOURIEZ.

employed in many sectors: petrochemicals, pharmacy, environment, food industry, energy, cosmetics, metallurgy, polymers, nuclear, water, inorganic, or organic chemistry.

4.1.1. *Prehistory of chemical engineering*

During the SFGP (French Chemical Engineering Society) Convention on Chemical Engineering held in Paris on November 4, 2016, Jacques Breysse recalled that, while chemical engineering is often presented as having started in the United States at the beginning of 20th Century (in particular with Arthur D. Little at MIT in 1915), then imported into France in the middle of the same century by Joseph Cathala in Toulouse and Maurice Letort in Nancy, we can cite great pioneers in France from the 19th Century:

– Antoine Lavoisier, chemist, philosopher, and economist, not only stated the conservation of matter as one of the fundamentals of chemical engineering, but also described, in 1793, some of what would be later called unit operations; he was also one of the first to combine experiments and mathematics for the study of chemistry;

– Jean-Antoine Chaptal, chemist, doctor of medicine, and politician, is known in particular for his work on improving the industrial production of hydrochloric acid, and was also the author, in 1807, of a work devoted to what was then called *Chemistry Applied to the Arts*, in which we find many of the classic chapter headings of a modern work on chemical engineering;

– Eugène Peclet, professor, who is presented as a physicist – unlike the previous ones – also described in his *Traité de la chaleur* (meaning Treaty of Heat), in 1843, several of our usual unit operations;

– Ernest Sorel, an industrialist, was a pioneer in the study of reactors in 1887 (well before the first edition of Octave Levenspiel's work), best known for his work on distillation at the end of the 19th Century;

– Anselme Payen, an industrialist, was a friend of Chaptal and the first chair of industrial chemistry at Cnam from 1839.

Videos

"De la chimie appliquée aux arts... au génie des procédés" (From chemistry applied to the arts... to chemical engineering) by Jacques Breysse:

https://youtu.be/JGaOmMwMxTI

4.1.2. *A crosscutting science serving society*

Besides the anteriority, it is especially important to remember from this historical reminder the crosscutting nature of this science, its close link with industry, and its societal dimension. Chemical engineering is, on the one hand, **an applied science**; this is the reason why efficiency, simplicity, operability, and agility are always sought, in view of industrial transposition, and, on the other hand, **an integrative science**, at the crossroads of chemistry, physics, mathematics, computer science, and increasingly biology: as many basic sciences on which chemical engineering is based and which it fuels. We can say with a touch of provocation that chemical engineering people are poor chemists, poor hydrodynamicians, pitiful heat physicists, very approximative mathematicians, but that they are (the only ones?) capable of operating a plant for transforming materials and energy in all its complexity, by combining all these disciplines.

It is often difficult to make decision-makers and the general public understand the diversity of the fields and professions concerned with chemical engineering. This diversity gives rise to the difficulty of providing key figures on the job market, annual turnover, or foreign balance, for example. However, it is a reality that, born from petrochemistry and industrial inorganic chemistry, chemical engineering has extended to the environment, energy, health, and food sectors: in other words, all the major societal challenges of the 21st Century! It was, moreover, to accompany this extension that "*génie chimique*" was called "*génie des procédés*" (chemical engineering) in French from the end of the 1980s. Today, we even want to use the methods of chemical engineering to study non-industrial systems: the human body, natural environments, and urban ecosystems, for example.

The same major classes of phenomena (reactions, material and energy transfers, hydrodynamics), the same methods, and the same paradigms (balances, unit operations, coupled transfers, systemic approach, modeling, etc.) stand in the face of this diversity of domains. This is what makes chemical engineering a science in its own right and gives people trained in chemical engineering the ability to move from one industry to another over the course of their career. Still in a crosscutting manner, ecodesign (in connection with the circular economy – see Boxes 2.2 and 2.5), intensification (for the miniaturization of equipment and the acceleration of phenomena – see Box 5.1), and product engineering (to design a production process based on the definition of the useful properties of a desired product – see Box 2.7) are developing today. And it is the crosscutting and applied approach of chemical engineering that makes these developments possible.

Finally, chemical engineering is basically **a science of scale changes**, because of its vocation for industrial transposition, therefore for a long time for up-scaling, and more recently downscaling (see Box 5.1), and also because to study such complex phenomena, one must be interested in what is happening at the molecular level, at that of interfaces, on the scale of devices, of manufacturing plants, and even complete industrial sites.

John Michael Prausnitz, in his *Danckwerts Memorial Lecture* in 2001, entitled "Chemical engineering and the postmodern world" (Prausnitz 2001), re-establishes the link between chemical engineering and society as follows:

> Chemical engineering also needs to be reinvented if it is to survive. [...] Because our unavoidable task is to serve society, our attitudes and our activities need to adjust to what a changing society expects and increasingly demands, even if these expectations and demands are not always consistent with our traditional beliefs.

The slogans of the last conferences of the discipline prove that the community has adopted Prausnitz's injunction to serve society, beyond the industrial world alone. **Thus, the industry must not only progress to increase its profits, but also contribute to the improvement of society by ensuring the well-being of citizens.** The modern version of this concern is found in the societal challenges of the environment, energy, health, and food.

4.1.3. *Chemistry, formulation, industrial chemistry, chemical engineering, and product engineering*

Chemistry is the science that studies the constituents of matter, their properties, transformations, and interactions. To distinguish it simply from chemical engineering, let's say that chemistry has to do – among other things – with small-scale manufacturing, typically in the laboratory, of a small quantity of product in order to study and characterize it. For this, there is no need to worry about the physical phenomena that will accompany industrial production, or to overly optimize manufacturing: these will be the missions of chemical engineering.

To produce a small quantity of aspirin (acetylsalicylic acid, according to its chemical name), a mixture of salicylic acid and acetic anhydride is stirred at around 50–60°C for 10 minutes in the presence of sulfuric acid (catalyst); then the acetylsalicylic acid is crystallized, filtered, and thus obtained in the form of crystals; it may be possible to then recrystallize them in order to obtain crystals of high purity (see Figure 4.1).

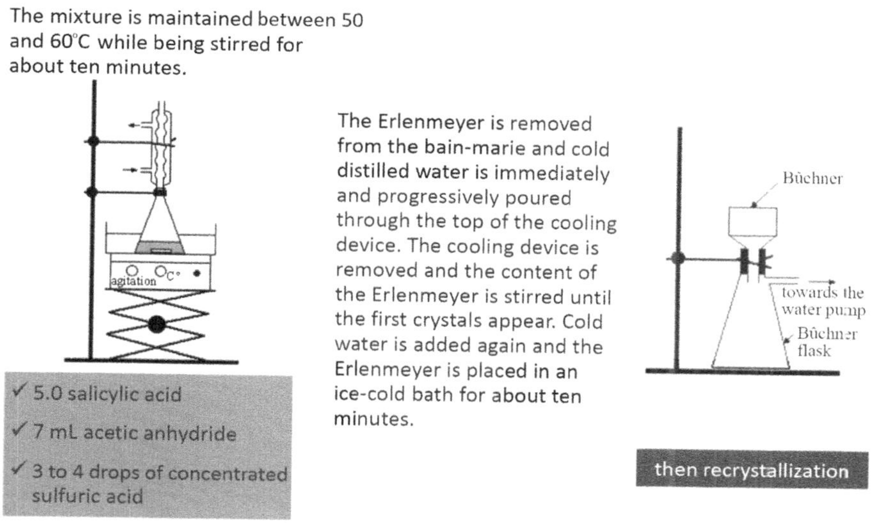

Figure 4.1. *Manufacturing aspirin: chemical version (source: (Mesplède 1995))*

Formulation consists of manufacturing – by mixing various substances – a product, homogeneous or not, with specific final properties that meet the requirements of a functional specification.

In the case of aspirin, it will be a question of mixing the active principle (acetylsalicylic acid) with an excipient in order to be able to manufacture tablets, or any other pharmaceutical form, guaranteeing a good conservation of the drug, the delivery of a precise dose of aspirin, and its correct release in the patient's body.

Chemical engineering was previously called *chimie industrielle* (industrial chemistry) in French. Today, the latter is the description of the production chains for major industrial products: petroleum derivatives, ammonia, sulfuric acid, etc. We can thus describe the ways in which aspirin can be produced from petroleum and natural gas, for example (see Figure 4.2).

Originally in France, ***génie des procédés*** was an extension of ***génie chimique*** to industries other than chemistry: pharmacy, the food industry, etc. Today, ***génie des procédés*** and ***génie chimique*** are synonymous.

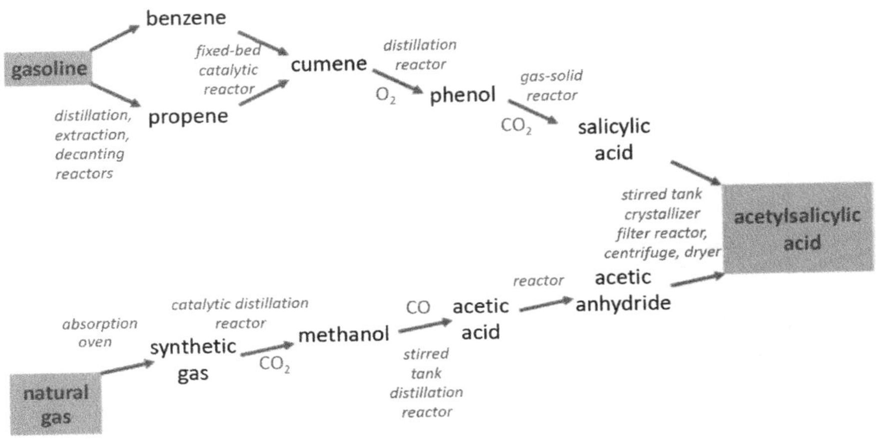

Figure 4.2. *Manufacturing aspirin: industrial chemistry version*

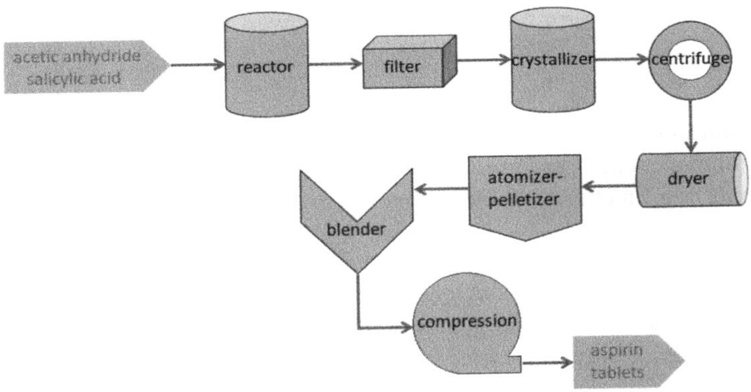

Figure 4.3. *Manufacturing aspirin: chemical engineering version*

Figure 4.3 shows the various stages of the industrial manufacturing of aspirin tablets: they are the essential stages already mentioned for manufacturing aspirin in a laboratory (see Figure 4.1), as well as other essential steps to guarantee a consistent quality of production of the finished product that is the aspirin tablet of a given dose.

Process engineering refers to a branch of chemical engineering dealing mostly with Modeling, Simulation, and Optimization of Processes (MSO).

Product engineering is a more recent science involving reverse engineering: designing the process leading not to the manufacture of one or more selected molecules, but to the industrial production of a product that has the desired functions (which is called a functional product) (see Chapters 2 and 3).

In the case of aspirin, it will be a question of producing antipyretic, analgesic, and anti-inflammatory drugs in an easy-to-swallow form (without predefining to its final formulation), releasing its active principle in a specific part of the body in a specific amount of time.

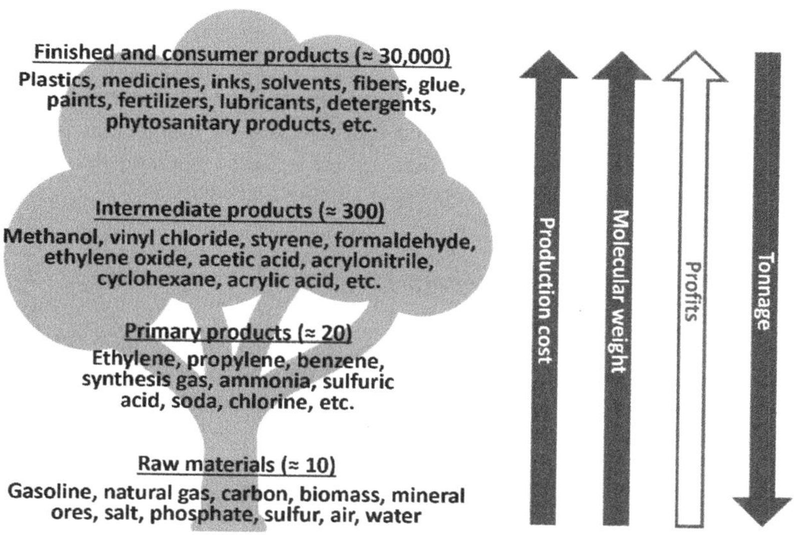

Figure 4.4. *Product tree (source: (Gani 2004))*

The **product tree** illustrated in Figure 4.4 shows that from a dozen raw materials, which is relatively invariant over the decades, we first produce about 20 basic products, which is also globally unchanged for a long time, followed by a few hundred intermediate products, and finally tens of thousands of finished products, the number and diversity of which continues to grow. From raw materials to finished products, the cost of production, molecular weight, and financial gains increase overall, while the number of metric tons produced per year decreases.

4.2. Fundamentals of chemical engineering

Basic sciences of chemical engineering are:

– thermodynamics, which contribute to the energy balances, the concept of equilibrium, either chemical or between phases, and its characterization by equilibrium constants and enthalpies;

– fluid (and powder) mechanics, which deal with the way material flows through pipes and equipment;

– heat and mass transfers, essential for any transformation;

– mass and energy balances, which are the very first tools to design processes;

– chemistry, mainly chemical kinetics and catalysis, the knowledge of which is necessary to design reactors.

Fluid mechanics, heat and mass transfers are often grouped under the single term "transport phenomena", as far as the first one is based on momentum transfer and balances too. We can even consider that, as far as these phenomena come under kinetics, they can be associated with chemical kinetics and catalysis.

4.2.1. *Thermodynamic fundamentals of chemical engineering*

The great adventure of thermodynamics – from the adjective *thermodynamics* that was proposed for the first time in 1849, from the Greek *thermos* (heat) and *dunamicos* (movement) – begins long before Carnot, Joule, Kelvin and Clausius, Gibbs and Duhem, even if its formalization is mainly the fruit of their work.

What exactly did the young polytechnician Carnot know?

Galileo (1564–1642) showed that a body remains immobile under the action of a resultant force of zero, but also that the speed of its fall under the action of the gravity force or "weight" does not depend on its quantity of the matter (as Aristotle believed) and increases over time. He thus laid out the foundations for statics and kinematics.

In *Principia*, Newton (1643–1727) defined, in 1686, the ("inertial") mass m of a body or quantity of matter as the product of its density by its volume and stated the three principles of what is now called classical or Newtonian mechanics:

– forces are the causes of the modifications of movement, but do not create it (Aristotle was wrong here too), and in the absence of forces, the natural state of a body is its movement at constant velocity vector v, not rest;

– its acceleration, or "increase of velocity", is inversely proportional to its mass m;

– the momentum or "impulse" or "linear momentum": product mv (vector) of a mass is conserved, but transformed.

It was known how to measure temperature indications, by observing relative heights of buoyancy level of ludions in a liquid (Galileo in 1592), of expansion of alcohol (Réaumur in 1730), then of mercury (Celsius in 1741; Linné in 1745). Thus, Celsius proposed a scale attributing the values 0°C and 100°C to the levels observed for the melting of ice and the boiling of water at atmospheric pressure. In 1783, Lavoisier showed that mass is preserved during a chemical transformation, but unfortunately considered that the heat released in combustion was a fluid without mass, calling it the "igneous" or "caloric" fluid (a fatal error for Carnot's reflections).

Figure 4.5. *Important figures of fluid mechanics, chemistry, and thermodynamics before Carnot and Joule. Portraits (from left to right) of Galileo Galilei (by D. Tintoretto), Isaac Newton (by Sir G. Kneller), Antoine Lavoisier (by J.L. David), Robert Boyle (painter unknown) (source: Wikipedia). For a color version of this figure, see www.iste.co.uk/dalpont/process1.zip*

His predecessors were interested in gases: Boyle and Mariotte (1662) observed that at constant temperature, the pressure P of a gas varied inversely to its volume V and that whatever the chemical nature of a given mass of gas, by choosing a temperature scale T such that $T = t_C + 273.15$, ***where*** t_C ***is the temperature in degrees Celsius***, the product PV divided by the mass m of the gas was proportional to T, which we call today the absolute temperature, in Kelvin. This proportionality was all the more true when the pressure was low. It was none other than our equation of state for ideal gases:

$$PV = \frac{m}{M_{Molar}} RT \text{ with } R = 8.314 \text{ J mol}^{-1} \text{ K}^{-1}$$

Finally, Charles and Gay-Lussac also showed, in 1787 and in 1802 respectively, that at constant volume V (isochoric conditions), the pressure P of a gas is proportional to T, regardless of its chemical nature.

At the time of James Watt (1736–1819) and Sadi Carnot (1796–1832), people sought to make “thermal” machines: the goal was to create as much movement as possible from a mass, using the heat input from the combustion of fossil materials (coal) to vaporize water: these machines are steam engines. First to operate coal mines, which requires pumping water from these mines and using this coal to make stronger steel; and also to transport, lift, dry anything: in short, the industrial revolution that we know.

From that moment, heat science and mechanics converged: there must be a single extensity in multiple forms – thermal, mechanical (potential, kinetic), and later chemical, electric (this will be energy, in Joule). Physics, which was until then a science of forces and momentum (hydraulics of medieval monasteries, Newtonian formalism) became the science of energy: energy is transformed, but its amount never changes, we could add, “from the Big Bang”. James Joule (1818–1889) demonstrated with the apparatus shown in Figure 4.7b, a reproduction of which can be found at the *Musée des Arts et Métiers* in Paris, that the heat and work provided to a system were flows of the same magnitude: “energy”. In documents from 1824–1832 that were discovered in 1878 after his death, Carnot already said: “heat is nothing other than motive power, or rather than movement that has changed shape. It is a movement in the particles of the body. Wherever there is destruction of motive power, there is at the same time heat production in an amount precisely proportional to the amount of motive power destroyed. Conversely: wherever there is production of heat, there is destruction of motive power” and “we can posit as a general thesis that motive power is an invariable quantity in nature; that it is never, strictly speaking, produced or destroyed. In truth, it changes form, that is to say that it sometimes produces one kind of movement, sometimes another, but that it is never annihilated.” But in his 1840 essay, *Dynamical Theory of Heat*, on Carnot’s (1840) famous ideal cycle heat engine, Joule concluded that only part of the heat supplied to the gas in his machine, when coming into contact with the hot tank, turns into work and that this gas must give off heat on contact with the cold tank. The conversion of heat energy into mechanical energy expected from the Carnot engine is not complete; its efficiency is limited. However, the work provided to a system can be entirely converted into heat – this is the case of cyclists’ energy that degrades into heat when they have to brake near a red light. Here, we can clearly see that a Joule of mechanical energy has a higher potential for valorization than that of a Joule of heat energy. Likewise, in the Joule device (Figure 4.7b), the water, as it cools, will not raise the weight that contributed energy in a mechanical state. Any transformation “degrades” energy.

Figure 4.6. *Players of the young thermodynamic discipline in the first half of the 19th Century. Portraits (from left to right) of Nicolas Léonard Sadi Carnot (by L.L. Boilly), James Watt (by C.F. von Breda), and James Prescott Joule (photographer unknown) (source: Wikipedia). For a color version of this figure, see www.iste.co.uk/dalpont/process1.zip*

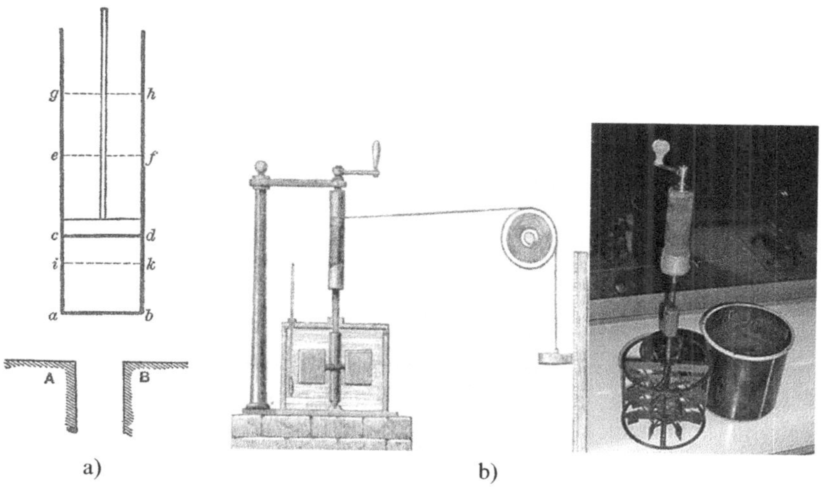

Figure 4.7. *a) Carnot's drawing of his ideal machine in his book Reflections on the Motive Power of Fire and on Machines Fitted to Develop that Power (originally published in French in 1824). b) Joule's apparatus (1869 engraving and creation by the* Musée des Arts et Métiers. *Source: Wikipedia). For a color version of this figure, see www.iste.co.uk/dalpont/process1.zip*

Thermodynamics emerged, followed by chemical thermodynamics: both deal with the transformations of energy from one form to another.

But let's take a short cut. Today, we deliver concepts of thermodynamics more or less abruptly and essentially mathematically. We talk about the state of matter, then we define state functions, the number of which has multiplied: internal energy, then enthalpy, entropy, free energy, free enthalpy, chemical potentials, and we "learn" principles like they were revelations. In truth, more humbly, these principles or postulates come from experimental observations and must be understood as follows: "all the tests carried out so far conform to such a statement", for example that of "the existence of an entropy function". Thus, thermodynamics focuses on the one hand on systems at equilibrium or "near-equilibrium", and on the other hand on systems where real physical processes take place.

Let us try to explain in a few lines the fundamental stages of thermodynamics, limiting ourselves to its two principles and to their use in very useful balances for chemical engineering.

4.2.1.1. *System, state, evolution, equilibrium, etc.: common words, but to be specified*

Thermodynamics first define what a system is: it is a material whole defined by its boundaries – a geometric surface, against the rest of the universe, called the exterior of the system. Thermodynamic systems are macroscopic, in the sense that they are supposed to pool together a large number of microscopic particles. This will, for example, be all of the water molecules that constitute a liter of water that you heat in a kettle for your tea party.

The macroscopic quantities or "state" properties of the content of a system are of two kinds:

– the extensive state quantities are additive: masses, volumes, number of moles;

– the intensive or local state quantities do not depend on the size of the system: these will be, among the most conventional, pressure, temperature, and the concentrations of chemical species, or even density.

Thermodynamics then postulate that this content can be totally represented by a finite number, or "variance" v, of independent intensive state properties, which, together with one extensive property, will totally characterize its state.

It is therefore possible, among the infinite set of state quantities, to choose some of them, called state variables, of which the other quantities are then "state functions".

Thermodynamics define the notion of evolution of a system or path from one state to another and that of equilibrium for which no state variable evolves over time.

Let's add a few notions: we have all experienced water in the form of ice, liquid, and steam (although the latter is hardly visible, except when it can be "seen" as bubbles within liquid water that is heated to prepare hard-boiled eggs for a picnic, and which disappears in the atmosphere of the kitchen). We will speak of "physical" states of matter, here in particular of water. We will also say that we usually experience the existence of matter – pure species or a mixture – under three types of "phases": solid phase, liquid phase, gaseous phase. Under atmospheric pressure, it is only the rise in temperature, therefore a higher "thermal" agitation, which leads water molecules to get rid of interactions or constraints between them, and to destabilize the structures of the liquid phase to form a gaseous phase. There is no such clear difference between the two physical states: a gas at high pressure is not so fundamentally different from a liquid. Better still, for matter made up of a pure species, beyond a "critical" pressure P_C and a "critical" temperature T_C that are specific to this species, the interactions between atoms or molecules of matter are such that the distinction between the liquid phase and the gaseous phase no longer exists: it is simply in the physical "fluid" state – also called "supercritical" – without there being a transition border between a gaseous state and a liquid state. Finally, a gas can be strongly ionized by an intense electromagnetic field, with the thermal agitation of the ions being such that the temperature becomes higher than 2000°C; this is the case of the corona around the sun. This is called the "plasma" state of matter. Moreover, if this ionized gas is made up of species with very different molecular mass, as the less massive ones move faster, it will in fact have two temperatures.

4.2.1.2. *Energy balance as a consequence of its conservation and transformation*

Thermodynamics postulate the concept and uniqueness of energy – in different forms – or total energy E_T carried by matter. It is measured in Joule (or kg m^2 s^{-2}). This total energy of the matter of a system is a function of its state: it is the sum of its macroscopic kinetic energy E_K, resulting from the summation of the kinetic energy of its volume elements, its potential energy E_P (for example the one gained by the pendulum of a clock at the highest point of its movement, in the gravitational field exerted by the Earth) and its "internal" energy U – "inside the body", according to Clausius. Actually, this internal energy is the sum of the microscopic energies carried by its constituent particles: kinetic (translation, vibration, rotation) and potential (stored in chemical or intra-atomic bonds, or in dipoles induced by force fields). U, E_K, and E_P are state functions and their variations from one state to another of a given mass of matter do not depend on the intermediate states through which it has passed, or on the "physical" forms of energy that we provided to it, most of the time heat (thermal energy) or work (mechanical energy).

The "first principle" simply states that energy, like matter and momentum, is neither created nor lost, but is transformed. The application that concerns the unit operations of chemical engineering is an instant balance of energy conservation (in Joule per second or Watt) for a system open to material streams: these streams not only carry their own ***total energy*** ($\dot{E}_T$), but also the "flow work rate" PQ (pressure × volumetric flow rate) associated with the volumetric displacement that they generate – it is also the mechanical power brought by these streams. The system is then the organ itself (dryer, reactor, crystallizer, etc.) where the unit operation is carried out. The balance is thus written as:

$$\dot{W}_{mec,walls} + P_E Q_E - P_S Q_S + \dot{q} = \dot{E}_{TS} - \dot{E}_{TE} + \frac{dE_T}{dt}$$

Here:

$$\dot{W}_{mec,walls} = \dot{W}_{Shaft} - P_{EXT}\frac{dV}{dt}$$

sums up the mechanical power supplied by the mobile walls of the system – turbine or compressor, if any, and the pressure on the liquid surface during filling for example; P_E and P_S (Pa) are the inlet and outlet pressures, Q_E and Q_S (m^3/s) the inlet and outlet flow rates, $\dot{q}$ (W) the thermal power supplied through the walls of the organ.

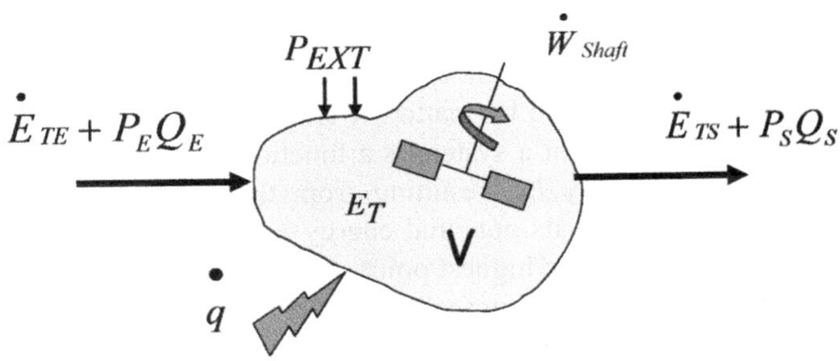

Figure 4.8. *Total energy balance for an open system*

The usual way to set up this balance from the more familiar one in a closed system is to consider a fixed mass of matter pushed into the system by one pipe and emerging from the other, much like an octopus marrying the geometry of the walls.

It should be noted that the above general energy balance does not provide any information on the distribution of total energy in the system as well as in the output material flow between internal energy, potential energy, and mechanical energy. In particular, if the system is a continuous open reactor, some of the thermal power supplied through to its walls can be used for the chemical transformation of the matter (in case of an endothermic reaction), while the rest effectively heats the fluid that goes through it.

Disregarding the fractions of macroscopic kinetic energy and potential energy in the system as well as in material flows, and assuming the pressure equilibrium $P_{EXT} = P_S$ (which is acceptable if the volume change of the system is slow compared to input and output material volume flows), the previous energy balance becomes:

$$\dot{q} + \dot{W}_{Shaft} = \dot{H}_S - \dot{H}_E + \frac{dU}{dt} + P\frac{dV}{dt}$$

where:

$$\dot{H} = \dot{U} + PQ$$

represents the sum of internal and mechanical energy flows, also known as enthalpy flow. For a continuous operation in steady state, the transient terms disappear:

$$\dot{q} + \dot{W}_{Shaft} = \dot{H}_S - \dot{H}_E$$

For a closed system without shaft work, we find:

$$\frac{dU}{dt} = \dot{q} - P\frac{dV}{dt}$$

where, with $H = U + PV$, the enthalpy content of the system:

$$\frac{dH}{dt} = \dot{q} + V\frac{dP}{dt}$$

The enthalpy state function H was originally introduced for transformations in a closed system with constant pressure, identical to the one of surroundings:

$$\frac{dH}{dt} = \dot{q}$$

The term enthalpy comes from the Greek word ευθαλπω – "bring heat inside" – with reference to that brought by the sun: during a transformation in a closed system under constant pressure, the variation of its enthalpy content is simply the thermal power provided.

We notice that the energy balance, whatever its formulation, involves finite (between input and output of a finite system) or elementary (temporal) variations of a state function: total energy, internal energy, enthalpy. It remains to be seen how these functions depend on state variables: pressure, temperature, nature, and mass or molar fractions of species. One of the most useful pieces of information is the mass heat capacity C_{Pi} (J K^{-1} kg^{-1}) of a species i: it is the amount of heat to be provided to a kilogram of this species to raise its temperature "at constant pressure" (a clarification that is especially important for a gas, because that of liquids and solids barely depends on pressure) by 1 K (Kelvin). Gay-Lussac (1806) and later Joule (1845) showed from experiments that variations in internal energy and mass enthalpy of a gas i with low pressure and temperature are proportional to those of its temperature: for enthalpy, the coefficient of proportionality is precisely C_{Pi}. This observation can be transposed to any gaseous, liquid, or solid species, except that C_{Pi} can itself depend on temperature. So, for example between the input and output of the system, the mass enthalpy variation of a species i is written as:

$$h_i(T_S) - h_i(T_E) = \int_{T_E}^{T_S} C_{Pi}(T)dT$$

Thus, for a system without chemical reaction or phase change, for example, it is enough to sum up the contributions of the Q_{mi} mass flows of all species after mixing in the feed to express the variation in enthalpic flow:

$$\dot{H}_S - \dot{H}_E = \sum_i Q_{mi}(h_{iS} - h_{iE}) = \sum_i Q_{mi} \int_{T_E}^{T_S} C_{Pi}(T)dT = (T_S - T_E)\sum_i Q_{mi} C_{Pi}$$

if these heat capacities vary little with temperature.

Let's go back to heat and work to conclude this section: what is their difference since, in previous energy balances, they have the same dimension and contribute in the same way to the variation of total energy in a system? In his work *Chaleur et Désordre* in 1999, Professor Atkins of the University of Cambridge formulated this difference very well: simply speaking, to work on a system is to impart a kinetic energy to its particles such that their velocity vectors are all parallel and identical: there is consistency in their movement; imparting heat means imparting a kinetic energy to these same particles such that their velocity vectors are all of different directions and intensity: their movement is incoherent or "disorderly", it is the well-known "Brownian motion" or thermal agitation, which, moreover, increases with temperature. Here, we see that work-energy must be of better quality than heat-energy.

4.2.1.3. *Isobaric, isochore, isothermal, adiabatic, reversible processes*

An isobaric transformation takes place at constant pressure, isochoric at constant volume, and isothermal at constant temperature. An adiabatic transformation is without intake or release of heat.

Thermodynamics introduce the theoretical and more delicate concept of reversible evolution or "conversion": it is constantly possible to reverse it by passing matter through the same states as those of direct conversion. As a result, this conversion must be a series of infinitely neighboring equilibrium states, thus "independent of time". These concepts are indeed seen in the mind: the content of a system is never rigorously at equilibrium, and our "actual" universe is the seat of irreversible spontaneous processes, whatever their characteristic time, changing the content of a system *towards* a state of equilibrium. However, this concept is of great interest, first of all to evaluate the changes in state functions of an actual conversion: we imagine a "reversible" process or conversion from the same initial state to the same final state. Above all, it quantifies the irreversibility of an "actual" transformation of matter, and calculates the evolution of the "quality" of the energy it carries during this conversion, that is its "entropy".

4.2.1.4. *Entropy*

It is beyond the scope of this work to explain more than a few elements of this fundamental concept, on the basis of a general formulation of chemical equilibrium relationships and between phases, and of the calculation of energy degradation in any actual kinetic process. Also, we encourage readers to refer to the excellent works listed in the upcoming section, "To learn more about thermodynamics".

Starting from the work on gases by Boyle, Charles, and Gay-Lussac, Carnot, and later Clausius, showed that heat and work inputs are not equivalent in "valorization" potential and led to this measurement of energy that cannot be recovered as "useful work", or the measurement of the level of energy degradation that is entropy. He imagines an engine to produce work from heat ("heat engine" – see Figure 4.8), with an ideal cycle made of four successive "reversible" conversions:

1) an isothermal expansion of gas at the temperature of a hot tank, thanks to a heat input;

2) an adiabatic expansion (that is to say without heat exchange) that brings it to the temperature of a cold tank;

3) an isothermal compression at this temperature that restores heat to the cold tank from the compression work provided (the colder temperature makes it possible to limit the increase in pressure);

4) an adiabatic recompression, which causes the gas to heat up to the temperature of the hot tank.

The work produced by his machine during the cycle is less than the heat imparted, since it was necessary to transfer heat to the cold tank: the engine is not one hundred percent efficient. If the system's content is an ideal gas, the equation of state of which has been presented above, it can be shown that the efficiency η of the machine work delivered/heat from the hot tank is simply:

$$\eta = 1 - \frac{T_F}{T_C}$$

where T, in Kelvin, is no longer a simple scale like the Celsius scale, but a thermodynamic quantity.

Beau de Rochas, Stirling and Diesel have devised other reversible cycles. Whatever the reversible cycle, any two equilibrium states connected by two different reversible paths $Tr1$ and $Tr2$ can be determined for this cycle, for which:

$$\int_{Tr1} \frac{\delta Q_{Rév}}{T_{EXT}} = \int_{Tr2} \frac{\delta Q_{Rév}}{T_{EXT}}$$

These evolutions are by definition those of the entropy function of the system, which is therefore expressed in J K^{-1}. The second principle is then stated as follows.

For a closed system, there is an extensive equilibrium state function S (J K^{-1}), called entropy, such that, if the system does not receive heat, its variation during any actual conversion – "irreversible" – be positive. As it is a state function, this variation is identical to that of a reversible conversion (for which then $T_{EXT} = T$) of identical initial and final states – that is the whole point of the concept of state functions:

$$S_2 - S_1 = \int_{TrRév} \frac{\delta Q}{T}$$

For example, the variation of entropy of an ideal gas in a closed system during an actual, that is to say an irreversible conversion, taking it from state (P_1,T_1) to state (P_2,T_2) is calculated by a reversible isobaric transformation from (P_1,T_1) to (P_1,T_2), followed by a reversible isothermal transformation from (P_1,T_2) to (P_2,T_2).

For a closed system receiving elementary amounts of heat δQ_j from tanks at temperatures T_{Pj}(K), the elementary variation of its entropy during any actual elementary transformation, from an equilibrium state, is the sum:

$$dS = \delta_E S + \delta_I S$$

with:

$$\delta_E S = \sum_{j=1}^{p} \frac{\delta Q_j}{T_{Pj}}$$

The variation of entropy $\delta_I S$ corresponds to irreversible internal processes, that is to say to all the physical and chemical kinetic processes that are discussed in the next chapter. Thus, for a reversible and therefore isentropic adiabatic process: $dS = 0$.

The absolute temperature T(K) thus becomes an intensive state variable independent from the measurement device considered. It is also called thermodynamic temperature, and is only a function of the measured temperature of a system. The entropy of a closed system whose T tends towards 0, itself tends towards 0.

The entropy balance at time t (J.s^{-1}K^{-1} or W K^{-1}) for an open system the fixed walls of which are maintained at T_0 is then written as:

$$\frac{\dot{q}}{T_0} + R_{\text{Pr}.Entr.} = \dot{S}_S - \dot{S}_E + \frac{dS}{dt}$$

What remains to be done is expressing the terms of entropy production $R_{Pr.entr}$, resulting from all transport and chemical reaction processes in the system, and the deviations from equilibrium that cause them: this is the objective of the thermodynamics of irreversible processes.

To conclude, thermodynamics address fundamental subjects for the chemical engineer: energy balances, equations of state of matter (gases, liquids, solids, supercritical fluids; pure or mixtures), phase and chemical equilibrium relationships, characterization of the interactions between species in a mixture, which reduce their "activities" compared to that expected from their concentrations, even minimization of the entropy generated (J K^{-1} kg^{-1}) by industrial production, that is to say a "technical-entropic" optimization in the same way as a technical-economic optimization (currency/kg).

To learn more about thermodynamics

de Hemptinne, J.C. (s.d.). Teaching of applied thermodynamics at IFP School [Online]. Available at: http://elan.ifp-school.com/RPG/ThermoAPP/story_html5.html.

Gicquel, R. (s.d.). "Thermoptim" Portal. École des mines de Paris [Online]. Available at: https://direns.mines-paristech.fr/Sites/Thopt/fr/co/conversion.html.

Schwartzentruber, J. (s.d.). Self-training module "Les bases de la thermodynamique : les principes fondamentaux et leurs applications directes". École des mines d'Albi [Online]. Available at: https://nte.mines-albi.fr/ThermoBase/co/ThermoBase_web.html.

Schwartzentruber, J. (s.d.). Self-training module "De la thermodynamique aux procédés : concepts et simulations". École des mines d'Albi [Online]. Available at: https://nte.mines-albi.fr/ThermoPro/co/ThermoPro_web.html.

4.2.2. Kinetic fundamentals of process design

"It's always tea-time".
Lewis Carroll
Alice in Wonderland, 1865

The industries that process matter (inorganic chemical, petrochemical, pharmaceutical industries, etc.) involve successive stages commonly called unit operations: gases are compressed and expanded, species are separated to keep only those of interest, reactive materials are carried and mixed, energy is supplied or extracted, most often in the form of heat, to carry out separations and reactions. In Chapter 5, several unit operations common to various industrial sectors will be presented, together with the calculations for their design. Their operation is generally based on transport phenomena, first on a local scale: heat transfer through a material, mass transfer in a liquid at rest, momentum transfer in a flow. They are described by laws based on experimental observations (laws of Fourier, Fick, and Newton). Chemical engineering then postulates that transfers to the contact surface between two phases – internal wall of the pipes of a heat exchanger, surface of air bubbles oxygenating a mechanically stirred fermentation reactor – can be described by the product of the difference in temperatures or concentrations, respectively away from and on the interface, and of a coefficient known as the "transfer coefficient". To calculate the dimensions of the equipment to be built for a specified product yield, chemical engineering has adopted suitable correlations, based on dimensionless numbers. These numbers are expressed on the basis of the parameters of the local laws cited above. It is thanks to their dimensionless nature, stemming from their physical meaning, that such correlations are relevant for the design of industrial size devices.

This section presents the local transfer laws previously cited, the concept of the transfer coefficient, the way those transfer correlations are written, and their similarity. Finally, to relate the efficiency of any unit operation to the design of the device that performs it, the chemical engineer must write mass, energy, and momentum balances on appropriate systems: we will illustrate this fundamental approach with two simple examples in section 4.2.5.

4.2.2.1. *Driving forces: the gradients at the basis of transfers*

Figure 4.9. *Portraits (from left to right) of Joseph Fourier (by L.L. Boilly), Adolf Fick (by M.A. Klamroth), and Isaac Newton (by Sir G. Kneller) (source: Wikipedia). For a color version of this figure, see www.iste.co.uk/dalpont/process1.zip*

4.2.2.1.1. Heat transfer by conduction: Fourier's law

Who has never been burned by touching a steel electric kettle and has not regretted the absence of a second and more insulating plastic layer? The difference between the temperature of the water that simmers inside and that in our kitchen, or that of our fingers, naturally causes a release of heat (we will call this a flow of thermal energy) to which the sensors of our epidermis are justifiably sensitive. But a plastic layer conducts much less heat and insulates better than a steel layer of the same thickness. In the 19th Century, Joseph Fourier (1768–1830; Figure 4.9) was able to establish from several experiments that the heat flow rate per unit area (or flux) J_T (J m^{-2} s^{-1}) in a direction $x'x$ perpendicular to a plane of a given material obeyed a phenomenological law (i.e. based on the observation of phenomena) proportional to the local temperature gradient T:

$$J_T = -\lambda \frac{dT}{dx}$$

where λ (J m^{-1} K^{-1} s^{-1}) is by definition the thermal conductivity of the material, that is to say its ability to conduct heat.

At temperatures between that of my fingers and that of the water in my kettle, the thermal conductivity of stainless steel is around 16 W m^{-1} K^{-1}, which is more than 20 times that of the polypropylene conventionally used to manufacture kitchen utensils. Hence our unpleasant sensation!

4.2.2.1.2. Mass transfer by conduction: Fick's law

Unless one wants to drink tea that holds barely any flavor, it is necessary to wait a few minutes for the tea from the sachet to infuse before tasting it. Even if the aromas of tea and theine are soluble in hot water, their transfer from the surface of the tea leaves into this water at rest depends on the Brownian motion of the water molecules at the temperature of water, which imposes its own dynamics.

In the 19th Century again, Adolf Fick (1829–1901; Figure 4.9) established, based on solubility tests of sodium chloride crystals and the analogy with Fourier's works and law, that the molecular flow rate of a species *i* diluted in a solvent per unit area (or flux) J_i (mol m^{-2} s^{-1}) in a direction $x'x$ perpendicular to the latter obeyed a phenomenological law proportional to the local gradient of the concentration of the species:

$$J_i = -D_i \frac{dC_i}{dx}$$

where D_i (m^2 s^{-1}) is, similarly to the thermal conductivity, a "mass" conductivity.

It is called the diffusivity or diffusion coefficient of the species in the solvent. This diffusivity depends mainly on the size of molecules of the species, the temperature, and the viscosity of the solvent. Its magnitude is usually in the range of 10^{-9} m^2 s^{-1} in a liquid solvent and 10^{-5} m^2 s^{-1} in a gas.

4.2.2.1.3. Momentum transfer by conduction: Newton's law

Another disadvantage of a stainless-steel kettle: its filter can be completely clogged with limescale. Tap water is not free from dissolved salts, even if water potabilization plants greatly reduce its hardness, particularly in limestone regions. At the boiling temperature of water, carbon dioxide, which is an acid gas, changes into a vapor: it is said to desorb, as opposed to absorb, and as the water becomes more basic, calcium carbonate crystallizes on the walls of the kettle and the filter. If water had no viscosity, clogging the filter would have no effect on the rate of flow out of my kettle when I pour water into my cup. But, instead, momentum flow decreases when passing through the dirty filter, because part of this momentum is transferred to the walls of the filter (due to the friction of water on the walls and on itself). The momentum flux J_{QM} (kg (m s^{-1}) m^{-2} s^{-1} or kg m^{-1} s^{-2}) towards the wall, in a direction $x'x$ perpendicular as in the filter, obeys a phenomenological law proportional to the local velocity gradient u_x. This is what Newton's viscosity law (1643–1727; Figure 4.9) states, from his work (1713) and Reynolds's tests (1883):

$$J_{QM} = -\eta \left(\frac{du_x}{dx} \right)_{wall}$$

The "momentum" conductivity η (Pa. s) is simply called the viscosity of the liquid, or sometimes absolute or shear viscosity.

The viscosity of water does not depend on flow velocity, which is a characteristic that defines so-called "Newtonian" fluids. Moreover, crystal deposits make the filter walls rougher, which increases the loss of momentum flow when water passes through the filter. As a consequence of the reduction of the flow cross-section due to limescale and, to a lesser extent, the roughness of the filter wall, the water flow over the same height will be smaller.

NOTE.– One can check that the unit of a momentum flux (flow rate per cross-section) is nothing but the one of a pressure (Pa). The friction on a wall therefore results in a pressure drop.

4.2.2.1.4. Kinetics of a reaction

A good soaking in vinegar should remove the limescale on the filter of my kettle, although this may take longer than with hydrochloric acid. Indeed, a classic descaling recipe involves soaking the object that has limescale in an aqueous acid solution. The dissolution that then takes place corresponds to the following chemical reaction:

$$CaCO_3(limescale) + 2H^+ \rightarrow Ca^{2+} + CO_2(gas) + H_2O$$

It is easy to see that the action of a strong acid, such as hydrochloric acid, for which all the hydrogen ions are available (the acid is said to be totally dissociated), will be much faster than that of a weak acid added with the same concentration, such as acetic acid CH_3COOH (which only partially dissociates in water). We'll say that the reaction rate, for example in g of $CaCO_3$ dissolved per m^2 of solid and per second, is greater with the first acid than with the second. According to kinetic experiments and data, it is proportional to the concentration of free hydrogen ions, which explains why acetic acid works more slowly than hydrochloric acid. Based on laboratory experiments, the expression of a reaction rate can be determined as a function of the reactants concentration and temperature (due to a higher frequency of collisions between reactant molecules by thermal agitation; this is called the activation of a reaction).

The driving force of a reaction is the deviation that the concentrations of reactants and products display from chemical equilibrium. Specialists of "thermodynamics of irreversible processes" would say that the affinity, based on the chemical potentials (in J/mol) of reactants and products, is positive. The rate of reaction is then the product of a pure kinetic term, related to a limiting kinetic step at molecular scale, and a thermodynamic term, which is for the deviation to the chemical equilibrium relation.

4.2.2.2. *Transfer coefficients*

4.2.2.2.1. Mass transfer coefficient

If I stir the contents of my cup, the infusion will be faster and my wait will be shorter. What is going on then? Stirring results in whirls and eddies that carry large volumes of solution away from the tea leaves. The transfer is greatly accelerated. Recording the concentration C_{im} of an aroma i in the tea, far from the surface of leaves, would show that its flux from the leaves to the tea, where it accumulates, is proportional to the difference between this concentration and the one C_{iS} at the surface of leaves, where its solubility is reached:

$$J_i = k_{Li}\left(C_{iS} - C_{im}\right)$$

k_{Li} (m s^{-1}) is by definition the transfer coefficient of species i in the liquid phase (subscript L) from the surface of the leaves. It will of course be all the greater as I give a higher mechanical power to my cup.

4.2.2.2.2. Heat transfer coefficient

But the stirring has other consequences: my cup becomes hot. The ceramic it is made of is indeed far from insulating. We feel that its temperature in our palm increases much more sharply than if we leave it still. In fact, here too, the same eddies carry large portions of fluid from the bulk of the tea towards the internal wall of the cup. The heat transfer to this wall is greatly accelerated. Recording the temperature T_m in the tea and the one T_P on the wall of the cup would show that the heat flux leaving the tea is proportional to their difference:

$$J_T = U\left(T_m - T_P\right)$$

where U (J m^{-2} K^{-1} s^{-1}) is by definition the heat transfer coefficient from the tea to the internal wall of the cup. Here too, it will be all the greater as I give a higher mechanical power to my cup. The temperature T_{EXT} of the external wall of the cup – the one I touch – is fortunately lower than that T_P of its internal wall. Back to the conduction of ceramics. In a sufficiently short period of time where the temperatures of the tea and of the internal and external walls of the cup are constant, the flux of thermal energy crossing the cup is constant throughout its thickness, not forgetting its radius of curvature; it has the Fourier's law expression presented upper. Assuming that the thermal conductivity of ceramics barely varies with temperature, it easily integrates:

$$J_T dx = -\lambda dT \rightarrow J_T \int_0^e dx = -\lambda \int_{T_P}^{T_{EXT}} dT \rightarrow J_T = \frac{\lambda}{e}(T_P - T_{EXT})$$

Thus, λ/e (J m^{-2} K^{-1} s^{-1}) is also *a* heat transfer coefficient, related to the transfer through the wall of the cup. Reconciling the two flux expressions:

$$J_T = U_T (T_m - T_{EXT})$$

where:

$$U_T = \frac{1}{\frac{1}{U} + \frac{e}{\lambda}} \quad (\text{J m}^{-2}\ \text{K}^{-1}\ \text{s}^{-1})$$

is an overall transfer coefficient between the tea and the outside wall of the cup; it decreases for a thicker cup. By analogy with the current through two electrical resistances in series due to a potential difference, it will be said that the heat flux crosses two transfer resistances in series, of respective values $1/U$ and e/λ.

4.2.2.2.3. Momentum transfer

By shaking the cup or by stirring, we give an impulse to its content: if we note u_m the average overall velocity gained by the tea alongside its wall, the momentum flow towards the wall, which has zero velocity, is proportional to the difference between the two speeds:

$$J_{QM} = k_{QM}(u_m - 0)$$

where k_{QM} (kg m^{-2} s^{-1}) is a "momentum transfer coefficient". We will see further on how this notion relates to the classic quantities of fluid mechanics.

4.2.2.2.4. Impact of the flow on transfer at an interface: calculation of transfer coefficients

Most often, it is thanks to correlations that the chemical engineer can estimate the values of mass or heat transfer coefficients that are essential for the *a priori* design of industrial-scale unit operation devices, that is to say higher than that of the devices used in a laboratory or a so-called "pilot" test hall. These correlations have been established for fixed geometry and proportions, based on numerous experiments, where the properties of fluids – density, viscosity, thermal conductivity, diffusivity – have been systematically varied. For instance, we can predict the coefficient of heat transfer to the wall of an industrial size cylindrical tank mixed by a turbine with six straight blades – thanks to the correlation proposed in the literature for the same proportions as in our equipment, according to a homothetic principle: same configuration, same ratio of the diameter of the turbine to that of the tank, same ratio of the height at which it is positioned relative to that of the surface of the liquid, etc. Unlike the correlations obtained from multiple linear regressions, those between dimensionless criteria are more relevant because these criteria compare quantities of physical meaning – characteristic times of kinetic processes, fluxes – of same unit. We only detail below the most important criterion in fluid mechanics, the Reynolds number.

Standard fluid versus laminar or turbulent flow

In a stagnant fluid, the chemical species only move from one point of the fluid to another by random motion, or thermal (Brownian) agitation. Thus, dye molecules injected into a point of a fluid (liquid or gas) will propagate throughout the fluid due to this thermal agitation: they "diffuse". If the dye molecules are injected at a temperature higher than that of the medium, the local thermal agitation or molecular kinetic energy will propagate both by their diffusion and also by their collisions with the molecules of the initial surrounding fluid. Thus, the propagation of heat in a fluid is always faster than the propagation of matter, which is the dye here.

Now, imagine a flow of fluid consisting of two parallel streams, each with a coherent overall motion, but two slightly different speeds, like two solid blades sliding one on the other: on the contact surface between the two streams, the elements of fluid belonging to the fastest stream let a part of their momentum (mass × velocity) to those of the slowest stream; they drag them along, all while staying close, without detachment between the two streams. This flow is said to be "laminar".

On the contrary, if the two preceding parallel streams have too different velocities, the viscous drag is no longer possible: the two streams detach one from the other and the velocity vectors of the fluid elements greatly differ in direction and in intensity. Fluid masses appear to rotate on themselves, the two fluids form vortices where they engulf one into the other and stretch – like a chocolate roll. The flow is said to be "turbulent". It is also the case of a fluid stream moving on the surface of a solid with a high differential velocity.

Let us take the classic example of the flow of an incompressible fluid, of density ρ kg m^{-3}), viscosity η (Pa s), sent with an average velocity u_m (m s^{-1}) in a cylindrical pipe of diameter L. The amount of macroscopic movement, called convection (from Latin *con-vehere*, to transport together), conveyed per unit volume of fluid is therefore $\rho.u_m$ (kg m^{-3}.m s^{-1}), while that transmitted by viscous friction to the wall, which has a zero velocity, is proportional to the ratio η/L also in (kg m s^{-1}).m^{-3}. The dimensionless Reynolds number is precisely the ratio of the first quantity to the second:

$$Re = \frac{\rho u_m}{(\eta / L)} = \frac{\rho u_m L}{\eta}$$

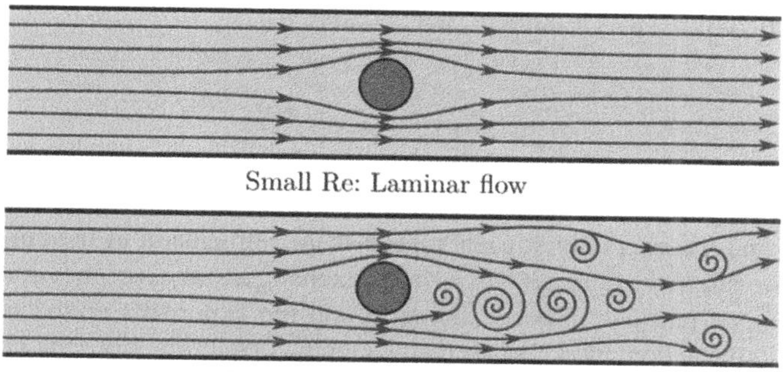

Figure 4.10. *From laminar flow to turbulent flow. For a color version of this figure, see www.iste.co.uk/dalpont/process1.zip*

It also allows another reading: the ratio L/u_m is a time that characterizes the transport of momentum by convection, while $\rho L^2/\eta$ is a time that characterizes that which occurs by viscous friction or "momentum diffusion":

$$Re = \frac{(\rho L^2 / \eta)}{(L / u_m)} = \frac{t_{Dif,QM}(s)}{t_{Conv}(s)}$$

Osborne Reynolds (1842–1912; Figure 4.11) showed through experiments that the flow in a smooth pipe changes from laminar to turbulent when $Re > 2000$.

Figure 4.11. *Portrait of Osborne Reynolds by John Collier, 1904 (source: Wikipedia). For a color version of this figure, see www.iste.co.uk/dalpont/process1.zip*

Relationship of transport phenomena to hydrodynamics – correlations

Most heat transfer correlations are in the following form:

$$Nu = a\,Re^{m}\,Pr^{n}$$

where Nu, called the Nusselt number, is expressed as:

$$Nu = \frac{UL}{\lambda}$$

U (W m^{-2} K^{-1}) is the heat transfer coefficient, λ (W m^{-1} K^{-1}) the thermal conductivity of the fluid, and L a characteristic or "macro-scale" dimension of the contactor – for example, the internal diameter of a heat exchanger tube or the diameter of a spherical catalyst particle. The coefficient a and the exponents m and n depend on the geometry of the contactor. When using such a correlation, note the choice, made by its author, for the dimension L: for a tank mixed by a mechanical stirrer, it may be the diameter of the tank or that of the stirrer. The Nusselt number is the ratio of a characteristic time of heat transfer by conduction over the distance L to that of transfer due to convection:

$$Nu = \frac{\left(\dfrac{L^2}{\lambda}\right)}{\left(\dfrac{L}{U}\right)} = \frac{t_{Cond,T}(s)}{t_{Tr}(s)}$$

Pr, called the Prandtl number, is expressed as:

$$Pr = \frac{\eta C_P}{\lambda}$$

As presented in Table 4.1, it is the ratio of momentum diffusivity to thermal diffusivity.

Likewise, most of the mass transfer correlations at an interface are presented in a similar form:

$$Sh = a\,Re^{m}\,Sc^{n}$$

Sh, called the Sherwood number, is expressed as:

$$Sh = \frac{k_{\phi i} L}{D_i}$$

$k_{\varphi i}$ (m s^{-1}) is the mass transfer coefficient of species i at the surface of phase φ, D_i (m^2 s^{-1}) the diffusivity of species i (which is also a material conductivity) in phase φ and L is still a characteristic dimension of the contactor.

The Sherwood number is also the ratio of a characteristic time of mass transfer by conduction over distance L to that of the transfer due to convection:

$$Sh = \frac{\left(\dfrac{L^2}{D_i}\right)}{\left(\dfrac{L}{k_{\phi i}}\right)} = \frac{t_{Cond,M}(s)}{t_{Tr}(s)}$$

Sc, called the Schmidt number, is expressed as:

$$Sc = \frac{\eta}{\rho D_i}$$

As shown in Table 4.1, it is the ratio of the momentum diffusivity to the diffusivity of species i.

Extensity	**Conductivities**	**Diffusivities (m^2/s)**	**Diffusivity relationships**
Species *i* (mole)	D_i (m^2 s^{-1})	D_i diffusion coefficient of i	$Sc_i = \dfrac{\nu}{D_i}$ Schmidt
Thermal energy (J)	λ (W m^{-1} K^{-1})	$\alpha = \dfrac{\lambda}{\rho C_P}$	$Pr = \dfrac{\nu}{\alpha}$ Prandtl
Momentum (kg m s^{-1})	η (Pa s) dynamic viscosity	$\nu = \dfrac{\eta}{\rho}$ kinematic viscosity	–

Table 4.1. *Conductivities of matter (Fick's law), heat (Fourier's law), and momentum (Newton's law) and corresponding diffusivities*

For the same contactor geometry, the heat and mass transfer correlations at an interface are often similar: this is because these two phenomena are governed by turbulence, characterized by the Reynolds number. For example, for a slightly turbulent flow ($4000 < Re < 10^5$) in a smooth tube, the reference work *Transport Phenomena* by Bird, Stewart, and Lightfoot (originally published in 1960) cites the two correlations:

$$Sh = 0.023\,Re^{0.8}\,Sc^{1/3} \text{ and } Nu = 0.023\,Re^{0.8}\,Pr^{1/3}$$

In conclusion, a whole series of correlations are provided in specialized works, such as the *Chemical Engineers Handbook* (CRC Press), according to flow geometries and polyphase contactor: plate, pipe, sphere, particles in packed beds, fluidized beds, moving beds, drops, bubbles in columns or agitated systems. Provided they are used under the conditions in which they were established, they are a very precious tool for designing industrial transfer and reaction devices. Currently another widely used method is the numerical simulation of flows (CFD for computational fluid dynamics), which makes it possible to test different geometries and to visualize maps of temperature and concentration of local velocities, but which requires good practice in defining flow meshes, in particular at the walls.

To learn more about transfer phenomena

Bacchin, P. (s.d.). Self-training module “Les bases des phénomènes de transfert”. Université de Toulouse [Online]. Available at: http://www.patricebacchin.fr/cours/phenomene_transport/co/transfert_web.html.

Bird, R.B., Stewart, W.E., Lightfoot, E.N. (2007). *Transport Phenomena*, 2nd edition. Wiley, New York.

Couderc, J.P., Gourdon, C., Liné, A. (2008). *Phénomènes de transfert en génie des procédés*. Lavoisier, Paris.

4.2.2.3. *Chemical kinetics: what reaction rate and mechanism?*

4.2.2.3.1. What reaction rate and mechanism?

In section 4.2.2.1.4, I had to wait a little while (a few minutes) to see the limescale that had built up in my kettle disappear completely, but I could easily see that it was disappearing. The action of an acid on limescale is a chemical reaction that is neither extremely slow nor instantaneous, but rather of an intermediate duration. However, I have heard that an explosion is an extremely rapid chemical reaction, although fortunately I have not experienced it myself. On the contrary, I know that the chemical reactions leading to the formation of petroleum are extremely slow and last for millions of years.

Knowing how quickly a thermodynamically possible chemical reaction takes place, that is to say for which the variation of free enthalpy is negative ($\Delta rG < 0$ – see section 4.2.1), is one of the objects of chemical kinetics. However, chemical kinetics is not limited to studying reaction rates and there are in fact other objectives:

– measuring and identifying the influence of various factors on the reaction rate (including temperature, as mentioned in section 4.2.2.1.4); understanding and controlling the influence of these factors is absolutely necessary for choosing the operating conditions to industrially implement a chemical reaction and in particular to design industrial reactors;

– establishing the rate law (a mathematical expression giving the speed of a chemical reaction);

– understanding the mechanism by which, at the molecular level, the rearrangement of chemical species takes place that conditions the reaction.

The vast majority of the systems that surround us seem inert on the scale of human time and for a chemical transformation to take place, in addition to being thermodynamically possible, is conditioned by an activation. An activation is a supply of energy to chemical reactants, which makes it possible to initiate the reaction. There are several ways to activate a chemical system: increasing the temperature (thermal activation), adding another constituent to the system (chemical activation), using radiation-like light waves (physical activation), or applying an electric current (electrochemical activation). For example, at room temperature, the $2H_2 + O_2$ gas mixture does not give rise to any visible reaction, whereas in the presence of a flame or a spark, it explodes.

What measure can we use to assess the reaction rate of a chemical system? For example, if I observe limescale dissolving in my kettle under the action of an acid, I instinctively perceive the pace of the reaction by watching the remaining amount of limescale that gradually decreases. Indeed, the reaction rate is given by the decrease in concentration of reactants (chemical compounds that disappear) according to time, or symmetrically, by the increase in time of concentration of products (compounds that appear). In the case of my kettle, I would obviously have a much harder time observing, for example, the appearance of Ca^{2+} ions with the naked eye because they are colorless. However, I observe the formation of small bubbles from an out-gassing of CO_2.

If we take the chemical equation of the reaction into consideration, the reaction rate is obtained from the instantaneous variations of concentrations relative to time:

$$v = \underbrace{-\frac{1}{\nu_A}\frac{dC_A}{dt} = -\frac{1}{\nu_B}\frac{dC_B}{dt}}_{\substack{\text{based on the speed} \\ \text{at which reactants disappear}}} = \underbrace{\frac{1}{\nu_M}\frac{dC_M}{dt} = \frac{1}{\nu_N}\frac{dC_N}{dt}}_{\substack{\text{based on the speed} \\ \text{at which products appear}}}$$

To measure the reaction rate, under given conditions, one observes the variation of a physical or chemical quantity whose value is related to the composition of the mixture. For example, in a gaseous system, when a reaction results in a change in the number of moles of gas, the change in total pressure can be tracked. The quantity to be tracked must be properly chosen and one is sometimes led to track the electrical conductivity of the reaction medium, its index of refraction, etc. The concentration of reactants or products in the reaction medium, whose evolution is shown in Figure 4.12, can also be tracked using chemical methods. To do this, small volume samples of the reaction medium are taken at regular intervals in which we quantify, via chemical titration, either a remaining reactant or a product that has appeared. This makes it possible to see the evolution of a concentration over time and thus to obtain the rate of the considered reaction.

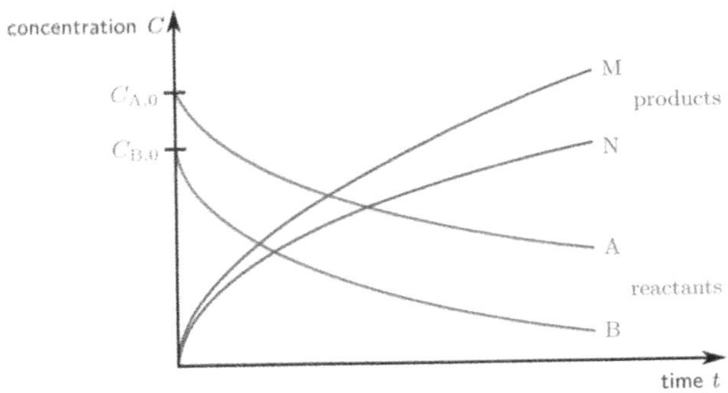

Figure 4.12. *Changes in concentrations of reactants and products over time. For a color version of this figure, see www.iste.co.uk/dalpont/process1.zip*

Depending on the concentrations of reactants, the reaction rate can be expressed as: $v = kC_A^{\alpha}C_B^{\beta}$.

k is a specific constant of the reaction studied. It is called the reaction rate constant. Exponents α and β are called reaction orders (with respect to A, B). These are usually positive or zero integers and sometimes fractional numbers. The sum $\alpha + \beta$

is called the overall reaction order. The reaction order is determined experimentally and cannot be known simply by examining the balanced equation of the reaction. However, the rate of some reactions does not obey a simple law of this kind.

There are two types of reaction systems, depending on the physical condition of the reactants and products. For example, in my kettle, the limescale is solid, while the other compounds are in liquid phase: the system is called heterogeneous because it has several phases. One can also encounter gas-liquid or solid-gas reaction systems. Unlike a heterogeneous system, a homogeneous reaction system only has one phase throughout its evolution.

The balanced equation of a reaction only provides information on the stoichiometry of the reaction, and does not reveal the intermediate steps leading from reactants to the products. All of these intermediate steps (which are intermediate reactions) together are called the reaction mechanism. There is a relationship between the reaction order and the reaction mechanism. Kinetic studies of chemical reactions therefore provide information on the reaction mechanism involved. Theoretical chemistry approaches, based on computer-based quantum chemistry calculations, can also reveal the reaction mechanism.

Identifying the physical or chemical factors that may or may not influence the reaction rate makes it possible to choose the most favorable operating conditions to carry out a chemical reaction. Indeed, many experimental conditions are likely to influence the reaction rate, for example:

– the physical state of reactants: in general, it is easier for two high-temperature gas molecules to meet (and thus react) than for a molecule of a liquid phase to diffuse through the reaction medium to a molecule present in solid phase;

– the concentration of reactants in solution or partial pressure of gases: the higher the concentration, the greater the probability of encounters between reactants;

– temperature: increasing thermal agitation also increases the likelihood of encounters;

– the presence of certain chemicals that do not appear in the chemical equation of reaction (for example, a catalyst, as will be seen in section 4.2.4.);

– for a heterogeneous system, the specific surface of a solid, for example, which explains the explosiveness of fine dust in grain silos.

At the industrial level, controlling all these parameters that influence the reaction rate is absolutely crucial to optimally conduct chemical reactions. Finding this optimum performance, under a large number of constraints (economic, environmental, etc.), is the challenge of chemical engineering. To summarize things in a simplistic way, it is first of all a question of foster encounters between reactants.

For this, chemical engineering has a multitude of levers at its disposal, in connection with the design of reactors adapted to the physical state of reactants and the operating conditions to carry out a reaction (temperature, pressure, concentration of reactants, time spent in the reactor, etc.). In addition to favoring encounters between reactants, it is also a question of facilitating their reaction after encounters, by lowering the activation energy necessary for the reaction: this is one of the challenges of catalysis, a key domain in process industries. Finding the optimum for industrially carrying out a chemical reaction makes it possible to address two highly interlinked challenges: economic efficiency and environmental compliance. For example, it may be unnecessary to heat an industrial reactor to 300°C if the reaction takes place almost as well at 100°C, as this represents both an energy and environmental cost.

4.2.2.3.2. Catalysis: accelerating and guiding a chemical reaction

What do the processes for making cheese, bread, beer, ammonia NH_3, sulfuric acid H_2SO_4, the exhaust after-treatment of your car, and so many other processes have in common? All these processes are based on chemical reactions that involve catalysts!

Among chemical reactions that are made possible by thermodynamics, most take place at speeds too low to be able to be carried out at the industrial stage. It is for this reason that more than 80% of industrial processes use catalysts to facilitate and guide chemical reactions.

A catalyst is a substance capable of increasing the reaction rate, without being consumed during the reaction, unlike a reactant. An industrial catalyst is selected to increase the reaction rate and even kinetically favor a reaction to the detriment of other undesirable reactions (those leading to unwanted by-products, in other words, wastes).

During a reaction between chemical compounds (reactants), chemical bonds are broken and others are reformed to lead to the formation of other compounds (reaction products). To break chemical bonds, it is necessary to provide energy, to cross an energy barrier. A commonly used image is that of a mountain pass separating two valleys (Figure 4.13). As seen above, there are several ways of supplying energy to reactants to enable them to cross this energy barrier and therefore react (this is called the activation energy of the reaction): we can, for example, increase the temperature or the pressure. Under certain operating conditions, the carrying out of a chemical reaction can then prove to be very energy-consuming; for example, increasing the temperature in an industrial-size chemical reactor represents a significant cost. It is in such situations that the use of a catalyst is advantageous, since it will make it possible to lower the energy barrier separating the reactants from the products. The reaction can therefore be carried out under milder conditions and an adequate catalyst can even make it possible to carry out a reaction with high selectivity, that is to say to favor obtaining one reaction product over other unwanted by-products. Indeed, a reactant can in general be involved in several competing reactions and it is then a question of favoring that leading to the targeted product.

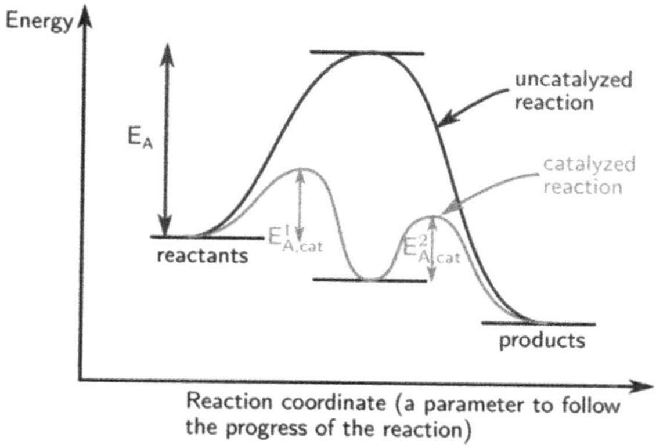

Figure 4.13. *Catalysis promotes a reaction by lowering the energy barrier: E_A is the activation energy of the uncatalyzed reaction; $E^1_{A,cat}$ and $E^2_{A,cat}$ are the activation energies associated with the catalyzed reaction. For a color version of this figure, see www.iste.co.uk/dalpont/process1.zip*

The activation energy of the uncatalyzed reaction, that is the height of the energy barrier, is greater than that used in the catalyzed reaction: with a catalyst, less energy must be supplied to the system to react. Even if catalysts generally contain very expensive metals, for example, platinum, their cost is largely amortized thanks to the energy savings that they make possible.

Catalysis therefore appears to be crucial in industrial processes, since it makes it possible to respond to both economic and environmental challenges: this ensures an ease in carrying out a chemical reaction and selectivity. Due to these elements, catalysis is one of the keys in developing sustainable processes, since it results in savings in terms of energy and raw materials, and limits waste generation during the reaction (non-recoverable by-products, to be eliminated or processed).

There are different types of catalysis, in particular depending on whether the catalyst is in the same physical state as the reactants. In homogeneous catalysis, the catalyst belongs to the same phase as the reactants, generally liquid. Enzyme catalysis is a special case of homogeneous catalysis, where proteins called enzymes play the role of the catalyst. For heterogeneous catalysis, the catalyst and the reactants belong to different phases: in general, a solid catalyst and reactants in the gas or liquid phase.

Industrially, in petroleum and petrochemical refining processes that represent large-scale productions, heterogeneous catalysis has largely supplanted homogeneous catalysis, mainly due to the ease of separating the catalyst and the reaction medium after reaction. Indeed, it is much easier to separate a solid catalyst from a liquid or gaseous mixture, than to separate a mixture of liquids, in the case of a liquid catalyst used in a liquid reaction medium. Heterogeneous catalysis is obviously used in fields other than the petroleum and petrochemical industry: for example, if the tea prepared in section 4.2.2.1 is rose-flavored, there is a good chance that it is due to adding 2-phenylethanol to tea leaves; and, this food flavoring can be industrially prepared using solid catalysts containing platinum.

Although not used frequently, homogeneous catalysis is used in certain fields of the industry, generally in the field of fine chemistry, for products with high added value that require complex organic syntheses. Because of its high selectivity, this type of catalysis is indeed widely used in the pharmaceutical field, that is to say for the chemical synthesis of active principles of drugs.

Apart from their importance in certain biological processes in living beings, enzymes are also used in industry, in particular pharmaceutical and food industries, to obtain certain products, and in the detergent industry to give certain properties to products (in particular ensuring the efficiency on different types of soiling). Black tea owes its taste and color to chemical reactions catalyzed by enzymes, which occur naturally after picking and handling tea leaves and the reaction pathway of which must be controlled. To make cheese, the first step in milk processing is coagulation, which is quite often carried out enzymatically. Enzymes are particularly useful due to their very high selectivity, since each of them is dedicated to a very specific type of reaction.

At the industrial level, the reactors that make it possible to carry out a heterogeneous catalysis are called catalytic reactors. There are several configurations for these reactors: fixed-bed, fluidized or suspended bed reactors, and structured reactors. In the fixed bed reactor, the reaction mixture percolates through a stack of catalyst pellets. In a fluidized or suspended bed reactor, small catalyst pellets are suspended in the reaction mixture. The structured reactor has a macrostructuration that channels the reaction mixture through channels or pores. We can, for example, find structures such as honeycomb, foam, or channels. More details on catalytic reactors will be given in section 5.5.

In fixed bed reactors, the catalyst is generally in the form of small pellets a few millimeters in size, either spherical or in other shapes. In the case of fluidized bed reactors, the catalyst pellets are very fine, a hundred microns in size. Finally, in structured reactors, the catalyst is fixed to the macrostructures constituting the reactor.

In heterogeneous catalysis, it is important to maximize the area of the interface between the catalyst and the reaction medium in order to accelerate the reaction rate. Indeed, it is at the surface of the catalyst that the reaction takes place and the larger the surface available to the catalyst, the more numerous the molecules of reactants that are capable of simultaneously approaching this surface, getting adsorbed therein, and reacting. One way of maximizing the contact surface between the catalyst and the reactants is to use porous solids, since reactants can penetrate and react inside catalyst pores.

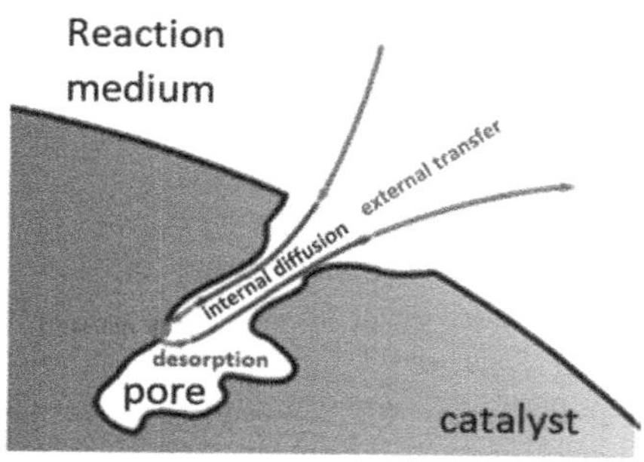

Figure 4.14. *Various stages of a heterogeneous catalytic process. For a color version of this figure, see www.iste.co.uk/dalpont/process1.zip*

A heterogeneous catalytic process can be broken down into different stages (Figure 4.14):

– diffusion of reactants in the reaction medium to the surface of the catalyst. This step comprises the transfer outside the catalyst and then diffusion inside catalyst pores;

– chemical adsorption of reactant(s) on the catalyst (chemical bonds are formed between the reactants and the catalyst, with the reactants "sticking" to the surface of the catalyst);

– reaction at the surface of the catalyst;

– desorption of products from the surface of the catalyst (the products obtained "break away");

– products are diffused into the reaction medium (internal diffusion followed by external transfer to the catalyst).

As these stages take place in succession, the entire catalytic reaction process takes on the speed of the slowest stage. The functioning of a catalytic reactor can therefore be limited by various kinds of phenomena depending on the case. In the case where the diffusion of reactants/products in the reaction medium takes place slowly compared to the stages involving chemical reactions (chemical adsorption of reactant, product formation, desorption of product), the reactor is said to be operating "physically" (since diffusion is an exclusively physical phenomenon). If one of the three chemical stages is the slowest in the catalytic process, the reactor is said to be operating "chemically".

Although there exist solid catalysts made of a single material (for example, finely divided metals such as Raney nickel, alloy of nickel and aluminum), solid catalysts generally consist of a non-reactive support, the catalyst material, and possibly a promoter. A promoter makes it possible to act on the catalytic activity, for example, by increasing its efficiency, its selectivity, or by prolonging its lifetime by giving it better thermal and mechanical stability.

For example, for the synthesis of ammonia NH_3 (the ammonia used for manufacturing fertilizers, plastics, etc.), an iron-based catalyst is used, containing various promoters that increase the catalytic activity of iron. For automotive depollution, platinum-, palladium-, and rhodium-based catalysts are used in the catalytic converters of petrol engines to eliminate carbon monoxide (CO), unburnt hydrocarbons, and nitrogen oxides (NO_x). These metal catalysts are deposited on a support made of magnesium silicoaluminate containing, in particular, alumina and rare earths.

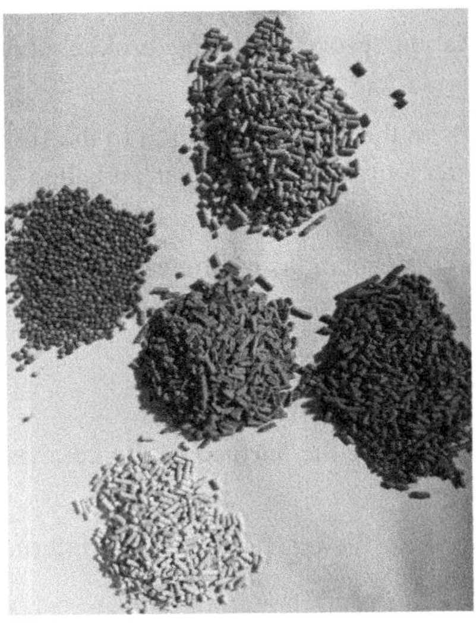

Figure 4.15. *Different types of catalysts (photo by Céline Houriez, 2019). For a color version of this figure, see www.iste.co.uk/dalpont/process1.zip*

By definition, a catalyst is not consumed during a chemical reaction. However, in practice, their lifetime is necessarily limited and the criterion defining this lifetime is called catalyst stability. When used in industrial reactors, catalysts can be "mishandled" by high temperatures, high pressures, corrosive environments, impact, etc. They end up worn out and losing their catalytic activity. They can also become fouled by the deposition of compounds like carbon that accumulate on their surface, clog the pores, and prevent reactants' access to the catalyst, or be "poisoned" by compounds that are very strongly adsorbing to their surface (like sulfur compounds on the surface of nickel catalysts).

4.2.3. *System-balances-performance approach for process design*

The task of a chemical engineer is to design and/or improve a production plant, with a well-defined and quantitative objective: most of the time, a plant is expected to deliver, at minimum unit cost (currency/metric ton), a given annual production (metric tons/year), in compliance with the market analysis, fulfilling QHSE laws and regulations (quality, health, safety, environment), and even, occasionally, to pack it in an adequate way for its customers.

4.2.3.1. *Example of a heat transfer device: water condenser in the secondary circuit of a nuclear power plant*

There are two main types of nuclear power plants: in what is known as the boiling water reactor (BWR), water vaporizes within the reactor core and the steam produced turns a turbine that generates an electric current. In what is called the pressurized water reactor (PWR, Figure 4.16), the liquid phase in contact with the reactor core exchanges heat with the water in a secondary circuit; it is then the resulting steam which makes the turbine turn.

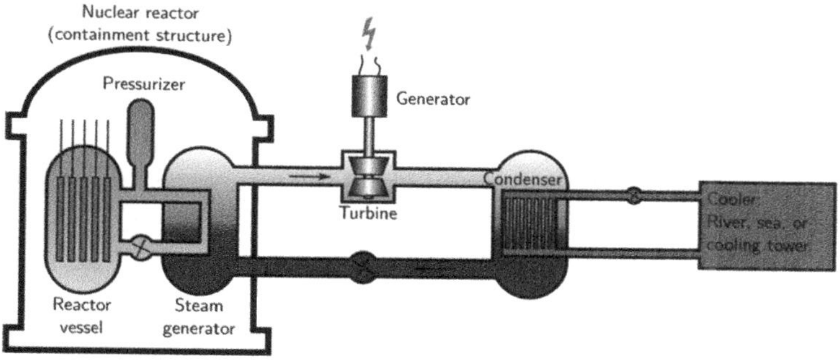

Figure 4.16. *Diagram of a pressurized water reactor (PWR) (source: Pâris Almageste, Wikipedia). For a color version of this figure, see www.iste.co.uk/dalpont/process1.zip*

The second type is obviously safer, because the primary circuit is entirely within the confinement structure. In both types of nuclear plant, the steam leaving the turbine is condensed before returning to the nuclear reactor. As an example, let us assume that the steam flow Q_{W_m} = 50,000 kg/s at a temperature T_W = 39°C, under a pressure of 69 mbar enters the condenser; this one consists of n = 100,000 parallel tubes, of internal diameter d = 1.2 cm, inside which river or sea water Q_{S_W} = 50 m³/h is pumped, with an input temperature T_E = 18°C.

Obviously, there is no question of returning the water from the secondary circuit without having completely liquefied it. The problem that the supplier of heat exchangers must solve is therefore the following: what power must be extracted from this steam to condense it? What then will be the temperature of the discharged river water (there are regulations for this and there is no question of scalding small fish)? And above all, what length L must the pipes be to attain this temperature?

From the pressure at which the steam flow reaches the condenser, the supplier concludes that it is practically that of steam at equilibrium with liquid water at 39°C: as soon as it enters, it begins to condense at the constant temperature of 39°C, as long as there are bubbles, owing to the heat transfer to the sea water circulating inside the bundle of pipes. To design the condenser, the supplier must know the enthalpy (or "heat") of condensation of the steam at 39°C: $\Delta H_{condensation} = -2{,}260$ kJ/kg. The negative sign of this enthalpy means that 1 kg of water in the gaseous state must lose energy to condense (conversely, a wet finger can identify which way the wind blows: the evaporation of the water and the transfer of the resulting steam are faster on the side of the finger which is windward, and the local energy consumption for evaporation causes a local drop in temperature, which we feel).

To condense all the steam $Q_{Wm}.\Delta H_{condensation} = +2{,}260$ MW must be evacuated. Where do we find this enthalpy, if not as an additional energy content of the river or the sea water leaving the condenser? At ambient temperature, the temperature of 1 kg of water increases by 1°C when it receives 4.186 kJ (it is also 1 kcal, kilocalorie): the corresponding quantity $Cp_W = 4{,}186$ kJ kg^{-1} K^{-1} is called the specific heat capacity of liquid water. And a m³ of liquid water corresponds to 1,000 kg: the density of liquid water ρ_W is 1,000 kg m^{-3}. The enthalpy conservation balance, on the system corresponding to the entire exchanger, is then written:

$$Q_{W_m} \Delta H_{liquefaction} = \rho_W C_{P_W} (T_S - T_E)$$

and hence, the river or sea water temperature at the output of the exchanger is 28°C, which is acceptable for a crocodile farm. What remains to be defined is the length of the tubes. The temperature of river or sea water in the tubes gradually increases due to heat transfer from the condensation of steam onto their external surface, but what is this temperature profile? Where it has a value T, the local heat transfer flux is, as we saw in section 4.2.2.2.2:

$$J_T = U_T\left(T_W - T\right)$$

where U_T (W m^{-2} K^{-1}) is an overall heat transfer coefficient including that of conduction through the thickness of the tubes.

Let us assume that, in a handbook for boilers and condenser design, we find as an average along the pipes $U_T = 3{,}000$ W m^{-2} K^{-1}. As the flux varies with T, the answer to the previous question can only be obtained by an energy conservation balance for a system with uniform temperature, which is here an elementary section of tube (see Figure 4.17):

$$0 = \rho_W C_{PW} \frac{Q_{SW}}{n}(T(z+dz) - T(z)) - U_T(T_W - T(z))\pi.d.dz$$

Or, after integration:

$$L = \frac{\rho_W C_{PW} Q_{SW}}{n.\pi.dU_T} \ln\left(\frac{T_W - T_E}{T_W - T_S}\right)$$

The length is 8.44 m.

Note that the last relationship can also be written as:

$$\tau = t_T \ln\left(\frac{T_W - T_E}{T_W - T_S}\right)$$

where:

– $\tau = \dfrac{n\pi d^2 L}{4Q_{SW}}$ is a time: it is the time taken by the cooling water to pass through the bundle of tubes of the exchanger, also called its "space-time";

– $t_T = \dfrac{\rho_W C_{PW} d}{4U_T}$, of course, is also a time: this is the characteristic time of the heat transfer to the walls of the exchanger.

The above performance relationship can thus be reformulated as follows:

$$\frac{T_S - T_E}{T_W - T_E} = 1 - \exp\left(-\frac{\tau}{t_T}\right)$$

The warming of seawater is all the more important, as the τ / t_T ratio is higher. It can be noted that its output temperature T_S is at most T_W, which can only be attained for an infinitely long space-time, that is to say a very low flow rate of sea water.

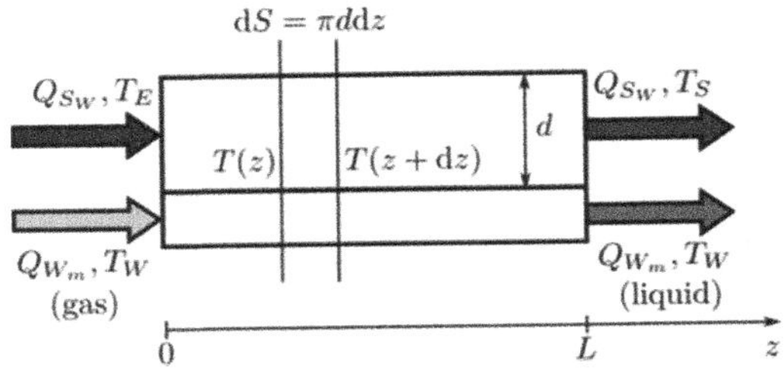

Figure 4.17. *Illustration of energy balance on an elementary section of an exchanger tube. For a color version of this figure, see www.iste.co.uk/dalpont/process1.zip*

4.2.3.2. *Example of a reactor: denitrification tank of a wastewater treatment unit*

A city wastewater biological treatment plant receives flows varying in quantity and composition throughout the day. Its role is, however, to discharge, into the natural environment (rivers), water that meets quality standards (regulatory limits of pathogen concentration, cations of heavy metals, putrescible oils and organic materials, temperature, etc. We can therefore imagine that it is strongly equipped with sensors (flow, temperature, composition, etc.), which themselves send signals to solenoid valves that tune the addition of reactants for water treatment, the air flow, even the flow supplied to the plant (the excess is temporarily placed in a “lagoon”). These “control-command” chains are also found in all industrial production facilities: products must comply with precise values of composition, whatever they are: petrol, polymers, tablets or pharmaceutical solutions, glasses, cosmetics, detergents, food products, and this despite the variability of raw materials (oils, ores, cereals, air gases, etc.) and even the variability of the temperature in the plant.

Let us imagine that a young chemical engineer is asked for a first sizing of a biological treatment unit meant to collect a volume flow rate of water $Q = 200\ \text{m}^3\ \text{h}^{-1}$, with a constant concentration of ammonium ions $C_A{}^E = 5\ \text{kg m}^{-3}$. In this unit ammonium ions degrade into nitrite ions, then into nitrates, and finally nitrogen, thanks to kind microorganisms (nitrosomonas, nitrobacter, etc.), in the presence of excess air bubbles in the water; this is called the aerobic fermentation process. It is assumed that a constant amount of microorganisms is maintained in the unit, due to a balance between their proliferation and their evacuation through attrition. The unit is to reduce the ammonium concentration to a required value $C_A^S = 0.5$ kg m^{-3}. How will the engineer calculate the volume V of the bioreactor for this objective?

Here are the two questions he/she absolutely must answer:

– How should the reactor be built? What geometry and volume should it have? How to achieve a good mixing and a good supply of microorganisms in oxygen?

– At which rate do bacteria consume the pollutant?

The first idea is to design a basin, stirred by a mechanical agitator, to better disperse the air flow necessary for bacterial life and ensure good heat exchange if necessary (bacteria are less active at low temperatures). For a sufficient stirring speed, the concentrations of dissolved oxygen, pollutants, bacteria, in short, the state of the liquid phase, in the thermodynamic sense of the term, are expected to be the same in all areas of the bioreactor. The engineer needs an assumption for the structure of the flow in the reactor, here that of a perfect "mixing", because the rate of reaction at a point in the basin depends on the local concentrations and temperature. Let us give the expression of the rate of consumption of the pollutant noted (kg of A. m^{-3}.h^{-1}) as a function of the pollutant concentration:

$$r = kC_A \; ; k = 2h^{-1}$$

The engineer can now translate what takes place in the reactor into a balance relationship: the pollutant is continuously consumed by the reaction in the bioreactor, so that its output is lower than its input. The assumption of a "perfect mixing" makes it possible to state that the reaction rate will be the same throughout the volume of the reactor. Furthermore, just as the liquid that is taken from a glass through a straw has the same composition as the liquid still in the glass, the water leaving the basin will have the same pollutant concentration as that in the basin. Then the balance that makes it possible to calculate the volume of basin necessary for the expected performances can therefore apply to the entire basin, and writes:

$$QC_{AE} = Q_{CAS} + r(C_{AS})V$$

With the reaction rate expression given, the engineer obtains:

$$V = \frac{C_{AE} - C_{AS}}{kC_{AS}} Q$$

or $V = 900\ \text{m}^3$.

It can be seen that the V/Q ratio is a time (s), which is here 4.5 hours. This is a fundamental characteristic of any continuous unit operation in stationary phase, here a bioreactor: its space-time, most often denoted as τ. We can also introduce the efficiency with which the reactor destroys the pollutant:

$$\eta = \frac{Flow\,of\;pollutant\;destroyed\,(kg/h)}{Maximum\;flow\,of\;pollutant\;destroyed\,(kg/h)} = \frac{QC_{AE} - QC_{AS}}{QC_{AE}} = \frac{C_{AE} - C_{AS}}{C_{AE}}$$

Thus, the material balance provides the relationship of efficiency of the perfectly mixed reactor to the quantity that characterizes its design, that is to say its space-time τ:

$$\eta = \frac{k\tau}{1 + k\tau}$$

Moreover, it depends on the product $k\tau$, which is dimensionless: the greater the kinetic constant k and/or the greater the space-time, the better the efficiency… which is not unexpected! On closer inspection: $k\tau = \frac{\tau}{\left(\frac{1}{k}\right)} = \frac{\tau}{t_C}$ is none other than the ratio of the characteristic time of the device to the characteristic time of the reaction: the greater the space-time compared to the reaction time, the greater the efficiency. This dimensionless product is called the Damköhler number. A similar characteristic reaction time can be obtained for any type of rate expression by normalizing it by its expression in a reference state (pressure, temperature, concentrations), for example that at the reactor's input.

Let us come back to our engineer: they are not satisfied with having to build a reactor of such a volume, when finally the reaction rate is very low everywhere, that is the one for the pollutant concentration at its output. What if the engineer puts three identical "perfectly mixed" reactors in series? The target concentration would only concern the volume of the last one and those in the previous ones would logically be higher. The first system to be considered is the first reactor. The pollutant balance is written as before:

$$QC_{AE} = Q_{CA1} + r(C_{A1})V \text{ then: } C_{A1} = \frac{C_{AE}}{1 + k\dfrac{V}{3}}$$

Similarly, the second system to consider is the second reactor and the pollutant balance will lead to:

$$C_{A2} = \frac{C_{A1}}{1 + k\dfrac{V}{3Q}} = \frac{C_{AE}}{\left(1 + k\dfrac{V}{3Q}\right)^2}$$

And for the third system, namely the third reactor:

$$C_{A3} = \frac{C_{A2}}{1 + k\dfrac{V}{3Q}} = \frac{C_{AE}}{\left(1 + k\dfrac{V}{3Q}\right)^3}$$

Finally:

$$C_{AS} = \frac{C_{AE}}{\left(1 + k\dfrac{\tau}{3}\right)^3}$$

The space-time required for the desired purification is now:

$$\tau = \frac{3}{k}\left(\left(\frac{C_{AE}}{C_{AS}}\right)^{1/3} - 1\right)$$

This space-time is 1.73 hours, and the total volume of the cascade is 346 m^3. It is much smaller, so is cheaper to build. But why then limit ourselves to three reactors? With n mixed reactors in series, we would get:

$$C_{AS} = \frac{C_{AE}}{\left(1 + k\dfrac{\tau}{n}\right)^n} \text{ and } \tau = \frac{n}{k}\left(\left(\frac{C_{AE}}{C_{AS}}\right)^{1/n} - 1\right)$$

And, remembering math lessons, for n tending to infinity:

$$C_{AS} = C_{AE} \exp(-k\tau) \quad \text{then: } \tau = \frac{1}{k} \ln\left(\frac{C_{AE}}{C_{AS}}\right)$$

Now the engineer finds a space-time of 1.15 hours, corresponding to a volume of 230 m^3. Perfect! But how can such an infinite number of reactors in series be built? Each of them should be of infinitely small volume since the total volume is finite. A sufficiently turbulent flow in a channel allows us to approach such a theoretical cascade.

This extreme situation corresponds to what chemical engineering calls a plug flow, referring to the image of an electrical outlet plug pushed into the wall socket. In French, this is called "*écoulement piston*", in reference to the image of a piston in one of the cylinders of a heat engine. Another image is that of wagons attached to each other moving on rails in a mine, or of a flow of ore on a conveyor belt. The principle is that fluid slices move one behind the other without mixing.

4.2.4. *Conclusion: ideal hydrodynamics and balances*

We finally defined two extreme ideal hydrodynamics:

– The system in perfect mixing, where the state (pressure, temperature, composition) is considered to be the same everywhere: the performances of a unit operation for which we can, as a first approximation, make this assumption, will be obtained from material, energy, or even momentum balances concerning its entire content. This is the case for processes where tanks provided with mechanical mixing are continuously fed and withdrawn as well as the operations in closed or semi-closed tanks, usually called batch and semi-batch operations (Figure 4.18; see also section 5.3). For instance, the pharmaceutical industry uses "fed-batch" reactors where a glucose solution is gradually added to the breeding medium in order to promote the metabolism of microorganisms which is in favor of antibodies. The best known example is the production of penicillin by fungi (penicillium).

– The plug-flow system where the state (pressure, temperature, composition) varies from the input to the output: the performance of a unit operation, for which we can, as a first approximation, make this assumption, will be obtained from material, energy, and even momentum balances on an elementary volume on the flow path.

Generally speaking, the energy balance (in J s^{-1}) is based on the first principle of thermodynamics written for open systems:

0 = incoming power – power entering + energy accumulation in the system

These powers (or energy flows) include:

– the internal energy and mechanical energy flows (pressure × volume flow rate) carried by the material flows entering and leaving the system;

– the thermal energy flows supplied through the walls of the system;

– the mechanical energy flows provided through the moveable walls – for example mechanical stirring – of a system.

Similarly, the balance on a species *i* (in mol/s, sometimes in kg of *i*) will simply show that its production by one or more reactions either accumulates in the system, or corresponds to a net output flow (output – input) of the system:

production of *i* = output flow of *i* – input flow of *i* + accumulation of *i*

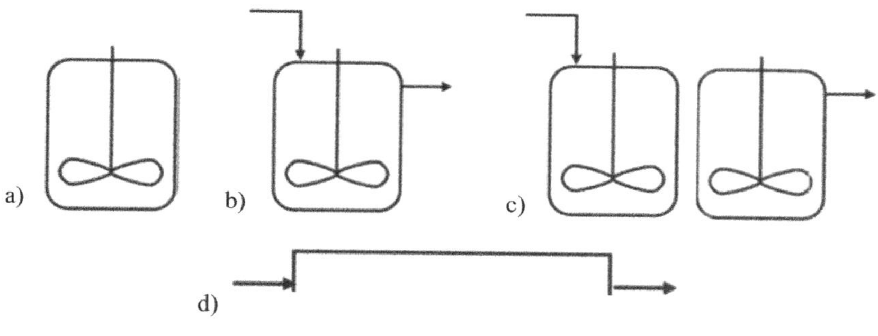

Figure 4.18. *(a) Closed stirred reactor; (b) continuous stirred reactor; (c) semi-closed or semi-open reactor; (d) plug-flow reactor*

See also section 5.5 and Box 4.1.

Most continuous unit operations use either tubes or columns, or open, semi-closed, or closed mixed tanks. To represent their deviation from ideal hydrodynamics presented, upper models are built based on the combination of perfectly mixed and plug-flow zones, with recycles, short-circuiting, etc.: for instance, the model of identical mixed zones in series, whose parameter is the number of mixed zones, makes it possible to represent a deviation from pure plug flow.

Here again, it is the solution of the balances on systems of uniform state (pressure, temperature, composition) that allow us to design and size a unit operation for a desired performance or tonnage.

4.3. Box

Writing material balances in chemical engineering

The conservation of material is written in the general case in mass: mass flows and mass. In some specific cases (constant density), we can consider writing them in volume (this is done in particular in hydraulics). Finally, during the study of reactors, we will switch to molar taking into account the fact that the stoichiometric coefficients refer to the number of moles.

Thus, for the typical process shown schematically in Figure 4.19, we will write the overall balance in general as:

$$Q_{m_1} + Q_{m_2} + Q_{m_3} = Q_{m_4} + Q_{m_5} + \frac{dm}{dt}$$

where Q_{m_i} are the total mass flows of the different flows i, m the mass of the system, and $\frac{dm}{dt}$ its variation depending on time (also called accumulation).

The balance during the same process for a particular species will be written as:

$$x_1 \cdot Q_{m_1} + x_2 \cdot Q_{m_2} + x_3 \cdot Q_{m_3} = x_4 \cdot Q_{m_4} + x_5 \cdot Q_{m_5} + \frac{dx \cdot m}{dt}$$

where x_i is the mass fraction of this species in flux number i.

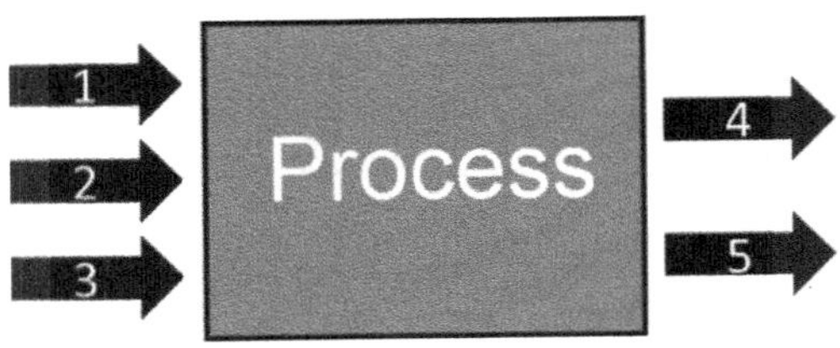

Figure 4.19. *Typical process: three inputs and two outputs*

To learn more about chemical engineering balances

Balice (Balances in Chemical Engineering) is a web app created by the ENSIC and the LRGP. For educational purposes, it can generate material, energy, and momentum balances: http://www.balice.fr/.

Videos

"Bilans de matière et d'énergie": a series of 12 videos, by Thierry Meyer (EPFL):

https://www.youtube.com/playlist?list=PLie7a1OUTSag0j5K57NYCIRbJ_eYYgLTd

Marie Debacq

Marie Debacq was a lecturer at Cnam, where her research work concerned the experimental study and modeling of multiphase reactors, in particular rotary kilns. Recently, she has been in charge of the technology platform at AgroParisTech.

Box 4.1. *Writing material balances in chemical engineering (Marie Debacq)*

4.4. References

Gani, R. (2004). Chemical product design: Challenges and opportunities. *Computers and Chemical Engineering*, 28, 2441–2457.

Marchio, D., Reboux, P. (2003). *Introduction aux transferts thermiques*. Presses des Mines, Paris.

Mesplède, J. (1995). *Chimie, terminale S*. Éditions ABC-Bréal, Paris.

Prausnitz, J.M. (2001). Chemical engineering and the postmodern world. *Chemical Engineering Science*, 56(12), 3627–3639.

5

Chemical Engineering: Unit Operations

The concept of unit operations simplifies/rationalizes the study of chemical engineering equipment in the face of the development of more and more technologies. Thus, the equipment that makes it possible to carry out the same type of operation is grouped together by a single term. These unit operations include: distillation, reaction, filtration, drying, etc. A production plant can therefore be divided into a succession of operations, each of which can be studied using unified tools and concepts. Conversely, it is possible to design a new manufacturing line, by modulating the unit operations necessary for that purpose.

The basic paradigms of the discipline are:

– material and energy balances (see section 4.2.5);

– combination of mechanical, thermal, and chemical phenomena (see sections 4.2.2 to 4.2.4);

– multi-scale approach (Figure 5.1). Indeed, we can envisage several scales of process studies:

- macroscopic scale: it is the human scale, that of things or phenomena that can be observed with the naked eye (even if it is sometimes through a camera),

- microscopic scale: this is the scale of films and particles, although the latter can have very different sizes. Let us say, for the sake of simplicity, that it is the scale of objects and phenomena around the μm,

Chapter written by Marie DEBACQ.

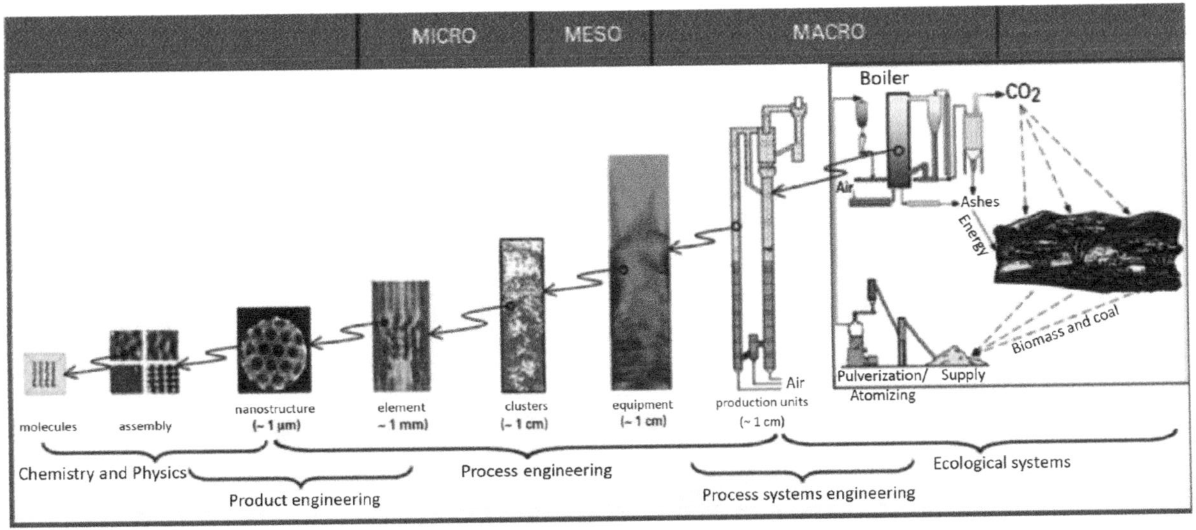

Figure 5.1. *Various hydrodynamic lengths/scales in the case of a fluidized bed reactor (Charpentier 2013). For a color version of this figure, see www.iste.co.uk/dalpont/process1.zip*

- between the two, the mesoscopic scale, which in chemical engineering involves looking inside an apparatus without going down to the μm,

- we can also mention the molecular scale (nm or below) and the metascale or megascale, which is that of manufacturing units, production sites or even companies;

– various types of modeling (see Box 5.2);

– the concept of theoretical stage (based on thermodynamics) and transfer unit (based on transport phenomena).

5.1. Distillation

Distillation is the most widely used separation operation in the industrial world. Although not very efficient from an energy point of view, it is simple and generally presents little risk, which is why its use is widespread. This involves separating the constituents from a homogeneous mixture (often liquid, but sometimes gaseous or as a liquid-vapor mixture); the operation uses coupled heat and material transfers.

Videos
"La distillation en 17 min" – Génie des procédés, Cnam: https://youtu.be/V5JSD7QXvHo

The **principle of distillation** is based on the formation of vapor that is rich in the most volatile compound when a multi-constituent mixture is heated; then, the condensation of this vapor, as shown in Figure 5.2 which represents a laboratory distiller: the mixture is heated in a round-bottom flask (2) using a heat source (1), which is a Bunsen burner here. The vapor that forms is richer in the most volatile compound than the initial mixture; it goes to the distillation head (3), is then condensed in the refrigerant (5) thanks to the circulation of cold water (counter-current between (6) and (7)). The distillate is collected in the flask (8). The system may be completed by a vacuum pump connected at (9) thanks to the adapter (10). We can easily imagine that it is possible to repeat the operation: distill the distillate again and thus enrich it even more with the most volatile compound and start again until the desired purity is obtained. The thermometer (4) makes it possible to follow the progress of the distillation when the boiling temperatures of the various compounds of the initial mixture are known, or better still, the bubble point and dew point curves of the mixture (see Figure 5.4, section 5.1.1).

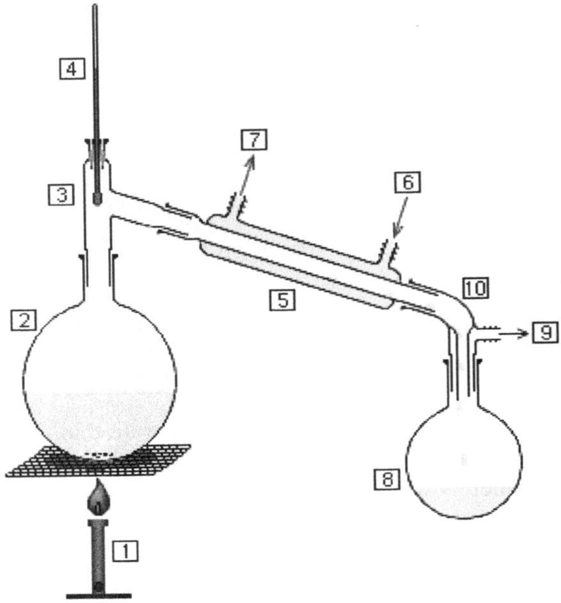

Figure 5.2. *Simple distillation often used in laboratories (source: H. Padleckas, Wikipedia, 2005). For a color version of this figure, see www.iste.co.uk/dalpont/process1.zip*

The **main industrial applications of distillation** are:

– the distillation of petroleum to obtain gasoline, kerosene, and diesel, as well as many other products that will then be supplied to various units of a refinery to form basic products of petrochemistry;

– cryogenic distillation of atmospheric air to produce the nitrogen and oxygen, necessary for many branches of industry (aeronautics, automotive, chemistry, electronics, metals, glass, etc.), medicine, food, pharmacy, etc.;

– the distillation of alcoholic drinks, generally after alcoholic fermentation of agricultural products.

When implemented on an industrial scale, flash distillation has only one stage and is therefore comparable, in principle, to a simple laboratory distiller. Strictly speaking, it should be called rectification when there are several stages, but in practice it is readily called distillation in all cases. Industrial distillation can be carried out continuously or discontinuously. Here, the focus will be on continuous distillation. It will also be limited in this case to binary distillation, that is the separation of only two components.

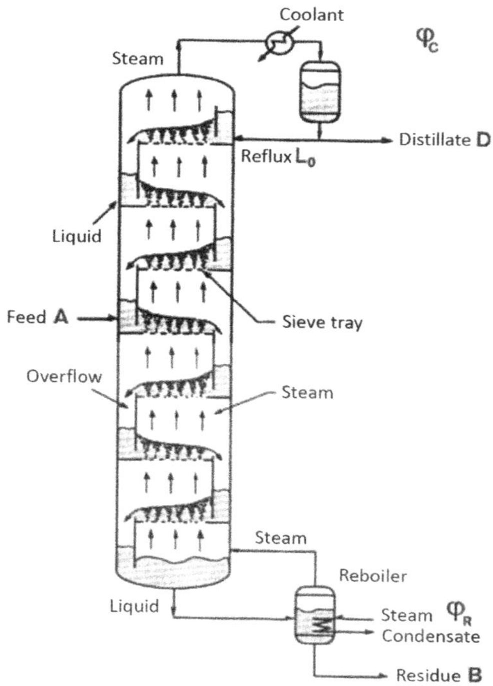

Figure 5.3. *Continuous industrial distillation (according to (Humphrey (2001)). For a color version of this figure, see www.iste.co.uk/dalpont/process1.zip*

Figure 5.3 shows the operating principle of a continuous industrial distilling column: liquid flows from top to bottom and steam from bottom to top; the material and heat exchange takes place throughout the height of the column between these two phases. As it ascends, the steam, which is created in the boiler (or reboiler) at the bottom of the column, is enriched with the most volatile compound; conversely, as it descends, due to gravity in the column, the liquid formed in the condenser at the over head of the column, and re-injected into the column at the reflux level, is enriched with the least volatile compound. The distillate is taken continuously at the top of the column, in liquid form, after the condenser. The residue (or bottom product) is taken continuously at the bottom of the column, also in liquid form, at the level of the boiler. The feed is injected in the middle of the column, at a level that must be determined precisely, in order to ensure that the column functions properly. The column section above the feed is called the rectifying (or enriching) section, while the column section below the feed is called the stripping section. In Figure 5.3, seven plates with a feed on the fourth plate are shown; later, we will see how to determine the number of plates necessary for a given separation, as well as the position of the feed.

Videos

"Distillation Basics: How a Distillation Column Works" – AIChE Academy:

https://youtu.be/M7AL7-44YTc

In order for the distillation to be carried out under good conditions, the following is recommended:

– a sufficient relative volatility α of the two constituents (preferably, $\alpha \approx \frac{P_1^{sat}}{P_2^{sat}} \geq 1,2$, where P_1^{sat} is the saturation vapor pressure of the most volatile compound and P_2^{sat} is the saturation vapor pressure of the least volatile compound);

– working with high flow rates (typically greater than 100 kg h^{-1}) and on chemically and thermally stable compounds, presenting no risk of explosion.

5.1.1. *Vapor–liquid equilibria*

As with any separation operation based on equilibria between phases, it is necessary to start by looking at thermodynamics (see section 4.2.1). Figure 5.4 shows how to represent vapor–liquid equilibria, with an isobaric diagram at the top and an equilibrium curve at the bottom, while emphasizing the correspondence between these two representations.

When a liquid mixture composed of the most volatile compound of concentration x_1 is heated, the first vapor bubble appears as soon as we cross the bubble curve (temperature T_1). The most volatile component has the composition y_1 in the first bubble. If we continue to heat the mixture (at temperature T_2) with an overall composition x_1, while remaining in the liquid-vapor zone, the most volatile product will have a concentration $x_2 < x_1$ in the liquid phase and a concentration y_2 in the vapor phase. And therefore, we will have $x_1 < y_2 < y_1$. As soon as we cross the dew curve (temperature T_3), the last drop of liquid disappears and the vapor returns to the composition with the most volatile compound $y_3 = x_1$.

Videos

"Temperature-Mole Fraction Diagram" – AIChE Academy:

https://youtu.be/lvRzkD-5Zyc

a)

b)

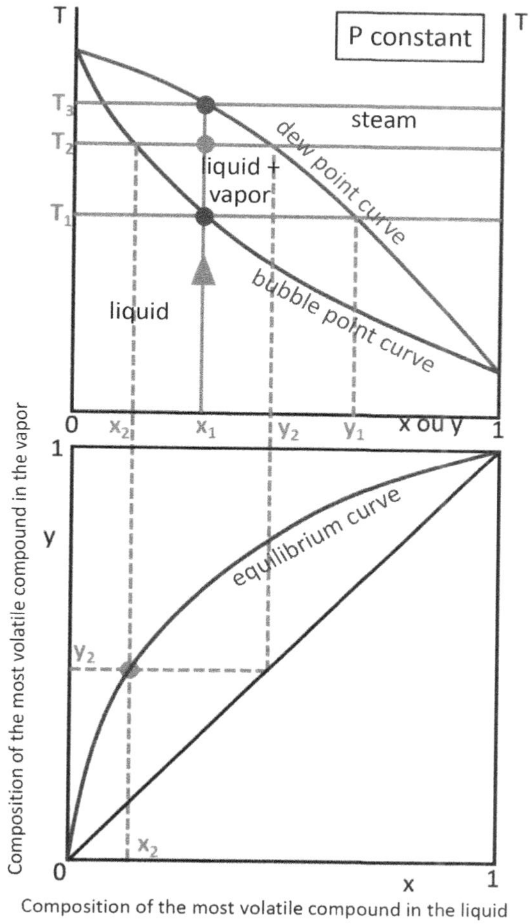

Figure 5.4. *Representation of vapor–liquid equilibria: a) isobaric diagram; b) equilibrium curve. For a color version of this figure, see www.iste.co.uk/dalpont/process1.zip*

For a given temperature T_2, the composition of the vapor–liquid equilibrium, composed of the most volatile compounds x_2 and y_2, respectively, can be read on the isobaric diagram. It is enough to transfer the point (x_2,y_2) on the equilibrium diagram to build the equilibrium curve point by point. We notice that this one passes:

– through (0,0): if the most volatile compound is no longer in the liquid phase, there is also none in the vapor at equilibrium with this liquid;

– through (1,1): if only the most volatile compound is present in the liquid phase, it is the same in the vapor at equilibrium with this liquid.

The equilibrium curve in Figure 5.4 is typical of an ideal system. There can also be mixtures with an azeotrope, that is a pinch between the bubble point curves and the dew point curves on the isobaric diagram, which results in the equilibrium curve crossing the first bisector and which significantly complicates distillation. This, for example, is the case of the water-ethanol mixture and this is the reason why – with conventional distillation – ethanol cannot be purified beyond 96%.

5.1.2. *Balances for a distillation column*

Balances for a continuously operating distillation column are written as follows:

– overall material balance: $A = B + D$;

– balance of the most volatile compound: $z_A \cdot A = x_B \cdot B + x_D \cdot D$;

– heat balance: $A \cdot h_A + \phi_R = D \cdot h_D + B \cdot h_B + \phi_C$;

where:

– A, B, and D are the total flows to the feed, the residue, and the distillate, respectively;

– z_A, x_B, and x_D the fractions with the most volatile compound for the feed, the residue and the distillate;

– h_A, h_B, and h_D the enthalpies to the feed, the residue, and the distillate;

– ϕ_R and ϕ_C the heat flows to be supplied to the reboiler and to be evacuated to the condenser.

The reflux rate R is defined, for a continuously operating distillation column, as the ratio between the liquid flow returned to the column after the condenser L_0 and the distillate flow produced D: $R = \dfrac{L_0}{D}$.

FOCUS.– *Mole or mass.*

As long as we keep a coherent system (total flows, composition of the most volatile compounds and enthalpies), it is possible to work with molar as well as mass quantities in distillation. Generally, the system of equilibrium data available is retained: thus, if the equilibrium curve is provided in mole fractions, will work with total molar flow rates and molar enthalpies.

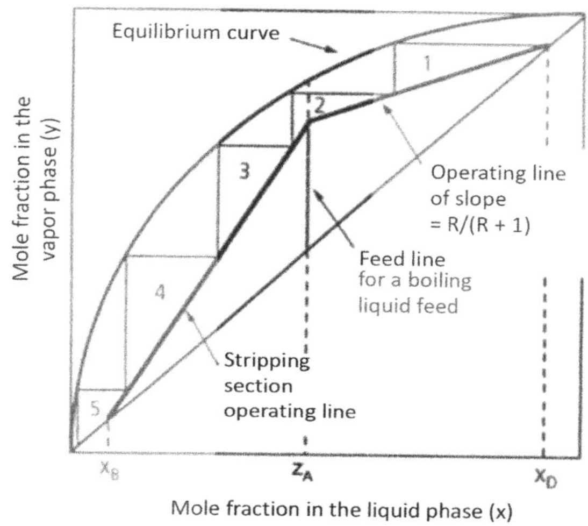

Figure 5.5. *McCabe–Thiele method (according to Humphrey (2001)). For a color version of this figure, see www.iste.co.uk/dalpont/process1.zip*

5.1.3. *McCabe–Thiele method*

The McCabe–Thiele method, illustrated in Figure 5.5, makes it possible to graphically determine the number of equilibrium stages (NES) of a distillation column:

1) We start by positioning, on the first bisector, the points corresponding to the fraction of the most volatile compound, x_D in the distillate, z_A in the feed, and x_B in the residue.

2) We then draw the rectifying section line, passing through (x_D, x_D), the slope of which is equal to $\frac{R}{R+1}$, and the feed line, passing through (z_A, z_A), the slope of which depends on the physical state of this feed (the feed line is, for example, vertical if the feed is in the form of boiling liquid, or horizontal if the feed is in the form of saturated vapor).

3) Then, we draw the stripping section line from the intersection between the two preceding lines and the point (x_B, x_B).

4) Finally, we draw the steps starting with (x_D, x_D). The number of steps is none other than the NES: the number of the stage that goes above the intersection between the rectifying, feed, and stripping section lines provides the position of the column's feed.

Videos
"McCabe-Thiele Graph Demonstration" – AIChE Academy: https://youtu.be/UNhlQyWYH4k

The McCabe–Thiele method is based on the assumptions of Lewis:

– the column is adiabatic, that is to say thermally insulated from the outside;

– the mixing heat of the two constituents of the binary mixture is zero;

– the vapor enthalpies of these two constituents are close.

Lewis's assumptions lead to the fact that the total liquid flow rate is not modified when crossing a plate (on the other hand, its composition changes); the same applies to the total vapor flow rate.

This remarkable simplification leads to the existence of operating lines (rectifying section and stripping section), which – without these assumptions – would not be lines.

FOCUS.– *Total reflux and minimum reflux.*

Total reflux → minimum NES; minimum reflux → infinite NES

At total reflux, the minimum number of equilibrium stages is obtained, but the distillate production is zero.

On the contrary, at the minimum reflux rate (therefore, for maximum distillate production), it would take an infinite number of stages to carry out the separation.

There are other methods for determining the NES of a distillation column. For example, we can cite the Ponchon-Savarit method that uses the enthalpy/composition diagram and makes it possible to dispense with Lewis's assumptions.

For industrial use, distillation columns are designed using specific, more or less sophisticated software, in particular modular simulators, which rely on thermodynamic databases (see Box 5.2). However, the usage of graphic methods makes it possible to understand the challenges of optimizing columns and thus to better control the behavior of the simulator.

5.1.4. *Technologies for continuous distillation*

Industrial distillation is carried out in columns (long vertical shells), that have either a packing material or equally spaced trays. In either case, the challenge is to promote the material and heat transfers between the liquid, descending due to gravity in the column, and the rising vapor.

> Videos
>
> 3D animation detailing the operation of a distillation column – Magenta Werbeagentur Mannheim:
>
> https://youtu.be/I70jgRpf80o

The **packed columns** have, over a few meters in height, either random packing (small objects thrown randomly into the column), or structured packing. These packings make it possible to increase the transfer surface between the gaseous and liquid phases, while leaving a sufficient degree of vacuum to allow for good circulation of the liquid and the vapor, without an excessive pressure drop. These packings are placed on support grids. At the head of the column, a distributor distributes the reflux liquid over the entire section of the column. When the column is very high, collectors and redistributors are placed between the packing sections in order to ensure good distribution of liquid, and therefore good contact with the rising vapor.

> Videos
>
> "Random Packing Demonstration" – AIChE Academy:
>
> https://youtu.be/yoWNau1Fqak
>
> "Structured Packing Demonstration" – AIChE Academy:
>
> https://youtu.be/p5S6Ri5aemA

To make a connection between the number of equilibrium stages (determined, for example, using the McCabe–Thiele method) and the height of the column, we use the concept of height equivalent to a theoretical plate (HETP): this is the height of the packing, corresponding to an equilibrium stage. It is a characteristic of the packing used, as well as of the type of distilled mixture.

Structured packings are known to be more effective than random packings; they are also easier to assemble/disassemble, less subject to compaction, and have a reproducible HETP. Random packings are less expensive than structured packings.

Plate columns have horizontal plates that are arranged at regular intervals. These are perforated plates, and each perforation may be fitted with a valve or a bubble cap. Vapor rises in the column through the perforations, possibly by raising the valves or caps. Liquid circulates on the surface of the perforated plates and passes from one plate to a lower plate via downcomers.

Videos

"Sieve Trays Demonstration" – AIChE Academy:

https://youtu.be/iahkOxbZ4Rk

"Valve Trays Demonstration" – AIChE Academy:

https://youtu.be/BdsM3bboeFM

"Bubble Cap Trays Demonstration" – AIChE Academy:

https://youtu.be/6_3HxK9ruOM

5.1.5. *Conclusion on distillation*

To pre-design a continuously operating distillation column, it is necessary to:

1) Graphically determine the minimum NES (at total reflux), to ensure that this number is reasonable and that the separation by distillation is well suited for the purpose.

2) Graphically determine the R_{min} (corresponding to an infinite NES).

3) Calculate the operating reflux rate (generally $R = 1.2$ to $1.5.R_{min}$).

4) Graphically determine the NES according to the McCabe–Thiele method described above.

5) Estimate the height of the column using the HETP (height equivalent to a theoretical plate) in the case of packed columns or based on the plate efficiency for column plates.

To fully design the distillation column, it is then necessary to focus on heat and mass transfers to precisely determine the height of the column, and on the hydrodynamics of the column, to calculate its diameter.

To learn more about distillation

Humphrey, J. (2001). *Procédés de séparation. Techniques, sélection, dimensionnement*. Dunod, Paris.

Schwartzentruber, J. (n.d.). Self-training module: "Distillation" [Online]. École des mines d'Albi. Available at: https://nte.mines-albi.fr/ThermoPro/co/Module_Distillation.html.

Seader, J.D. (2006). *Separation Process Principles*. Wiley, New York.

Treybal, R.E. (1981). *Mass Transfer Operations*. McGraw-Hill, New York.

5.2. Fluid–solid mechanical separations

In a manufacturing plant, a frequent problem is that of separating a pulverulent solid from a fluid: it is known as suspension when the fluid is a liquid and smoke when the fluid is a gas.

To carry out this separation, purely mechanical operations such as decantation due to gravity, centrifugation, or filtration may be used.

Very schematically, decantation will be used for low concentration suspensions of small particles (1 to 10 μm), filtration for the separation of concentrated suspensions of particles of various sizes (5 to 1,000 μm), and centrifugation for intermediate cases: medium concentration and medium-sized particles (10 to 500 μm).

5.2.1. *Fluid–solid interaction laws*

Before being able to approach fluid–solid mechanical separations themselves, it is necessary to understand some concepts and equations of physical hydrodynamics.

Videos
"Les interactions fluide/solide : quelques lois utiles pour le génie des procédés" – Génie des procédés, Cnam: https://youtu.be/ROjQjFaKeb0

From the study of the flow of fluids around particles came the notions of:

– **particle Reynolds number**: Reynolds number for which the density ρ_f and viscosity μ are those of the fluid, the characteristic dimension d_p is that of the particles, and the characteristic velocity u is the relative velocity of the fluid with respect to the solid: $Re_p = \dfrac{\rho_f \cdot u \cdot d_p}{\mu}$;

– **drag**: T resulting from pressure and friction forces on a particle in the direction of flow: $T = C_x \cdot \dfrac{\rho_f \cdot u^2}{2} \cdot \Omega_c$. The drag coefficient C_x can be obtained in the literature depending on the shape of the particle (see, for example, Figure 5.6) and the flow regime; Ω_c is the projected surface of the particle in the direction of flow.

From the study of the fall of a particle in an immobile fluid, the expressions from Table 5.1 could be drawn, making it possible to calculate the fall velocity u_{t0} of an isolated particle with diameter d_p and density ρ_p in a fluid with density ρ_f and viscosity μ, depending on the Archimedes number:

$$Ar = \frac{d_p^3 \cdot g \cdot (\rho_p - \rho_f) \cdot \rho_f}{\mu^2}$$

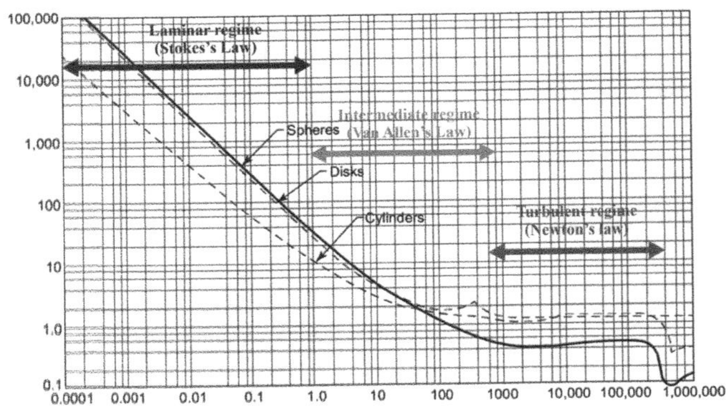

Figure 5.6. *Drag coefficient (y axis) depending on the particle Reynolds number (x axis) (according to Perry and Green (1997)). For a color version of this figure, see www.iste.co.uk/dalpont/process1.zip*

Laminar regime (Stokes's law)	**Intermediate regime (Van Allen's law)**	**Turbulent regime (Newton's law)**
$Ar < 27.6$	$27.6 < Ar < 4.4 10^5$	$4.4 10^5 < Ar < 1.1 10^{11}$
$Re_{t0} = \frac{Ar}{18}$	$Re_{t0} = 0.153 \cdot Ar^{0.714}$	$Re_{t0} = \sqrt{3 \cdot Ar}$
$u_{t0} = \frac{d_p^2 \cdot g \cdot (\rho_p - \rho_f)}{18 \cdot \mu}$	–	$u_{t0} = \sqrt{\frac{3 \cdot d_p \cdot g \cdot (\rho_p - \rho_f)}{\rho_f}}$

Table 5.1. *Reynolds number and fall velocity (the particle Reynolds number as calculated at the fall velocity and is expressed as:* $Re_{t0} = \frac{\rho_f \cdot u_{t0} \cdot d_p}{\mu}$*)*

FOCUS.– *Non-isolated particle.*

As an initial approximation for a swarm of particles, the fall velocity u_t can be related to the fall velocity u_{t0} of an isolated particle, depending on the relationship $u_t = u_{t0} \cdot (1-\phi)^n$, where ϕ is the volume fraction of solid in the suspension.

The value of n can be determined using the Richardson-Zaki correlations:

$$n = 4.65 \text{ for } Re_{t0} < 0.2$$

$$n = 4.4 \cdot Re_{t0}^{-0.03} \text{ for } 0.2 < Re_{t0} < 1$$

$$n = 4.4 \cdot Re_{t0}^{-0.1} \text{ for } 1 < Re_{t0} < 500$$

$$n = 2.4 \text{ for } Re_{t0} > 500$$

Note that the presence of particles in a fluid modifies the density and the viscosity of the formed suspension as follows:

$$\rho = \phi \cdot \rho_p + (1-\phi) \cdot \rho_f \;\; ; \;\; \mu = \mu_f \cdot 10^{1.82 \cdot \phi}$$

In the case of a solid subjected to a centrifugal force, the centrifugation speed is obtained after evaluating the centrifugal effect $K = \frac{\omega^2 \cdot r}{g}$ (ω being the rotation speed and r the distance from the axis of rotation):

– $u_c = K \cdot u_t$ under the laminar regime (Stokes's law);

– $u_c = K^{2/3} \cdot u_t$ under the intermediate regime (Van Allen's law);

– $u_c = \sqrt{K} \cdot u_t$ under the turbulent regime (Newton's law).

The study of the flow of fluids in porous media leads to the following expressions for evaluating the pressure drop ΔP in a porous or granular medium with a thickness of L for a flow area Ω :

– Darcy's law: $Q_v = \frac{B}{\mu} \cdot \Omega \cdot \frac{\Delta P}{L}$, where the coefficient B is called permeability. The common unit for this is Darcy, which corresponds to 10^{-12} m^2; for sand, the permeability is about 100 Darcy; a cigarette, 1,000 Darcy; soil, 1 Darcy; limestone, 10^{-2} Darcy;

– Ergun's law: $\frac{\Delta P}{L} = h_K \cdot a_p^2 \cdot \frac{(1-\varepsilon)^2}{\varepsilon^3} \cdot \mu \cdot u_s + h_B \cdot a_p^2 \cdot \frac{1-\varepsilon}{\varepsilon^3} \cdot \rho_f \cdot u_s^2$, the first part of which corresponds to the Carman-Kozeny law (for low Reynolds numbers) and the second part to the Burke-Plummer law (for high Reynolds numbers); ε is the porosity of the porous medium, that is to say the fraction of volume not occupied by the solid (note that this porosity varies depending on the state of the settled bed); a_p is the area per unit volume of the particles (for spherical particles, $a_p = \frac{6}{d_p}$); generally $h_K \approx 4.2$ and $h_B \approx 0.3$, but one can find other values or expressions in the literature, according to the type and form of grains considered.

The surface velocity $u_s = \frac{Q_v}{\Omega}$ of the fluid is the ratio of the volume flow rate over the flow area, that is to say the velocity calculated on the cross section of the container as if it were empty. The average velocity of the fluid in the pores $\overline{u_p}$ is necessarily greater than the surface velocity, since the actual volume of the fluid is lower, due to the presence of the solid: $\overline{u_p} = \frac{u_s}{\varepsilon}$.

5.2.2. *Settling*

The simplest devices are tanks of varying heights. The clear liquid is generally evacuated by overflow and the sludge is drained from the bottom. To facilitate sludge removal, it is most often necessary to install a mechanical scraper system. Tanks can be parallelepiped, cylindrical, or cylindroconical (Figure 5.7). These are large structures and are most often operated continuously.

To limit the volume of the settling tank, tubes or lamella can be used to improve the performance of the sedimentation basin; a lamellar settling tank is thus obtained. They can be seen, in particular, in wastewater or industrial water treatment operations or as a potabilization stage.

In the case where the fall of a particle can be considered as not influenced by the presence of neighboring particles and walls, the design of a settling tank is very simple: the comparison of the fall time $t_{fall} = \frac{H}{u_{t0}}$ of a particle isolated at height H of the settling tank and the time of passage $\tau = \frac{V}{Q_v}$ taken for the suspension to pass through the settling tank, with volume V that operates continuously, while being crossed by a volume flow Q_v makes it possible to determine whether this particle will undergo decantation (when $t_{fall} < \tau$) or entrainment (if $t_{fall} > \tau$).

Thus, the flow rate that a settling tank can treat is equal to the product of its surface area on the ground and the fall velocity of particles. A lamellar settling tank with n plates of dimensions $L \times l$ inclined at an angle α to the horizontal can treat a maximum flow $Q_{vmax} = n \cdot L \cdot l \cdot cos\alpha \cdot u_{t0}$; we then understand the space savings there will be with such a settling tank.

In the case of aggregated suspensions, designing settling tanks requires laboratory tests, and the use of calculation methods, based on the graphical analysis of the results of these tests (Kynch theory, in particular), make it possible to roughly design a settling tank. Today, research work makes it possible to develop more elaborate and more precise designing methods.

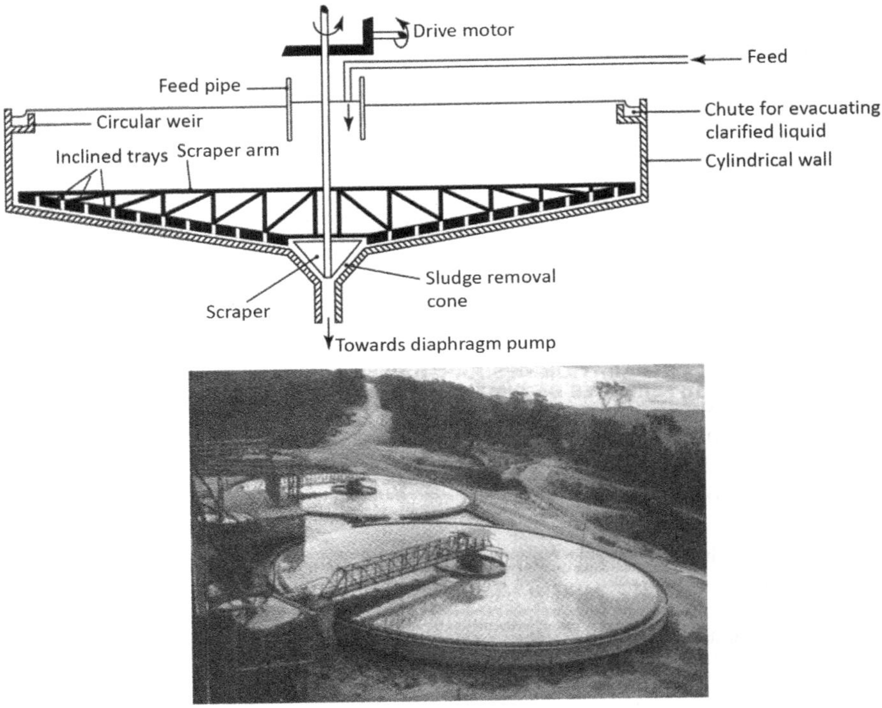

Figure 5.7. *Example of a circular gravity settling tank (source: Anglaret (1998)). For a color version of this figure, see www.iste.co.uk/dalpont/process1.zip*

5.2.3. *Centrifugal and inertial separation*

5.2.3.1. *Centrifugal decantation*

Centrifugal decantation brings us back to decantation by gravity by replacing g with $\omega^2 \cdot r$ (ω being the frequency of rotation and r the distance from the axis of rotation). It is faster than decantation by gravity; it will be used for:

– stopping a physicochemical or biological transformation;

– making process units more compact;

– cases where pressure and temperature must be controlled;

– products with high added value.

Centrifugal decantation can be used if the density of the fluid and of the particles are sufficiently different (relative deviation greater than 10%), the kinematic viscosity of the fluid is not too high ($< 1.5\ 10^{-4}\ m^2\ s^{-1}$), and particles are not too small (> 0.5 µm).

Centrifugal decantation is widely used in the food industry (skimming milk, wort separation in beer, clarification of fruit juices), for the recovery of catalysts and bacteria or even the separation of latex.

Among moderate-speed centrifuges (centrifugal effect $K < 3{,}000$), mention may be made to the simple bowl centrifuge and the raclette centrifuge. High-speed centrifuges ($K > 3{,}000$) are used to treat suspensions of very fine particles (around 1 µm): either in a tubular centrifuge or in a disk centrifuge (Figure 5.8), in which the bowl can have a diameter of 15 cm to 1 m and rotate up to 12,000 revolutions/min, to treat a flow rate up to 100 $m^3\ h^{-1}$ with a volume fraction of solid not exceeding 15%.

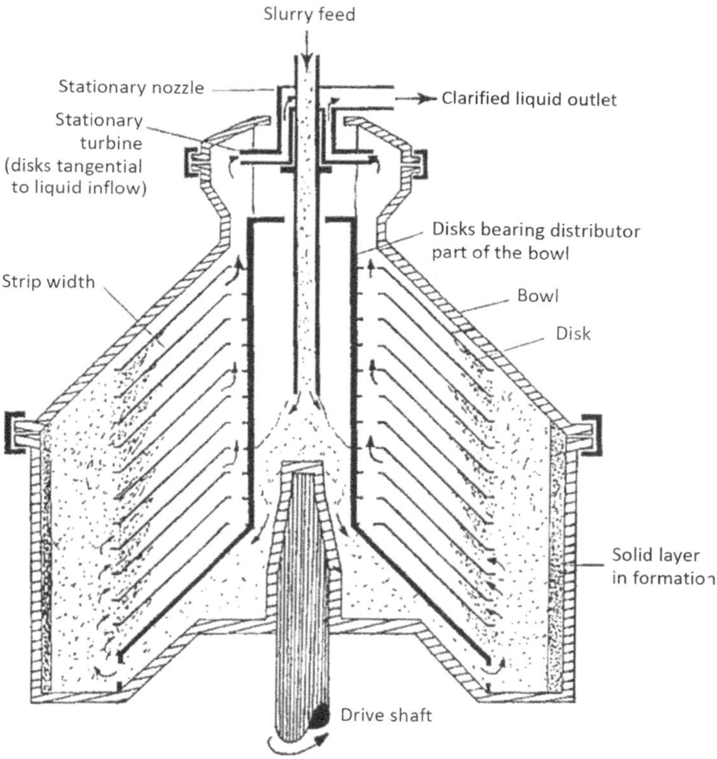

Figure 5.8. *Schematic diagram of a disk centrifuge (source: Anglaret (1998))*

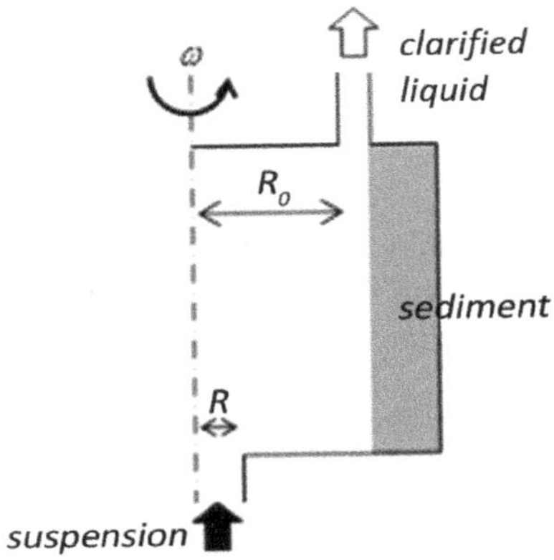

Figure 5.9. *Notations for a centrifuge*

Thus, the maximum allowable flow in a decanter centrifuge is $Q_v = \Sigma \cdot u_t$, with $\Sigma = \dfrac{\omega^2 \cdot V}{g \cdot ln\left(\dfrac{R_0}{R}\right)}$, where V is the filling volume of the centrifuge, R_0 the position of the evacuation of the clear liquid, in relation to the axis of rotation, and R the radius of the input nozzle of the suspension to be treated at the same axis of rotation (see notations in Figure 5.9).

Videos
"Three-phase centrifugal separator" – Andritz Separation: https://youtu.be/sDhueH3q6Y4

5.2.3.2. *Cyclones*

Cyclones are used mainly for gas-solid separation (cyclones) and more rarely for liquid-solid separation (hydrocyclones) of small particles (1 to 100 µm). They are simple structures (no moving parts) and involve very low operating costs. Finally, they make it possible to work at high temperatures (up to 1,000°C) and/or pressure (up to 500 bar).

They are widely used for particulate removal (to purify a gas or recover the solid), but can also be used for demisting (retrieving droplets).

Most cyclones have a tangential inlet (with velocities of a few tens of meters per second); evacuations are axial (Figure 5.10). To ensure a cyclone operates smoothly, various authors recommend precise proportions, which can be found in specialist books.

Videos

"Cyclone Separators Demonstration" – AIChE Academy:

https://youtu.be/2pm54ayF3TA

"High Efficiency Cyclones" – Hurricane System:

https://youtu.be/BrGXXurZers

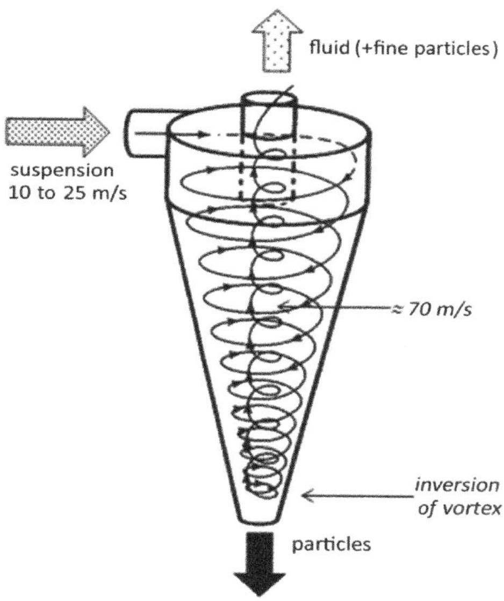

Figure 5.10. *Schematic diagram of a cyclone (according to: McCabe and Smith (2001))*

5.2.4. *Filtration*

Filtration separates fluid from the particles it contains faster than sedimentation (especially for small particles), but it is more expensive. There are several categories of filtration, depending on the size of particles, as shown in Figure 5.11.

The objective of filtration is to mechanically separate a continuous fluid phase from a pulverulent dispersed phase. Filtration involves passing the suspension through an adequate filtering medium, capable of retaining particles by physical action. We can carry out a bulk filtration or a deep-bed filtration (particle bed, for example): the particles gradually clog the medium. This technique will only be used if we do not want to recover the particles (for example, water treatment) and for suspensions low in particles of size 0.3 to 2.5 mm. It is also possible to filter on a medium with a large number of holes: the particles settle on the surface of the medium and will cake. This technique is used for more loaded suspensions and/or suspensions with smaller particles and/or if we want to retrieve the solid.

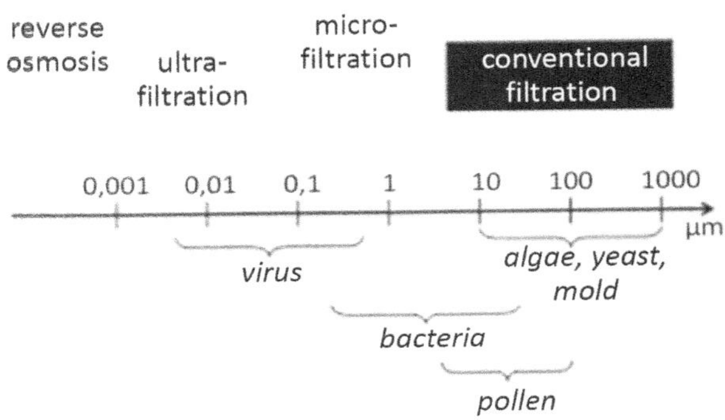

Figure 5.11. *Filtration categories by particle size*

Filtration through a mechanical support is most often a discontinuous operation, even if there are continuous filtration apparatuses (different zones of the apparatus will be reserved for successive phases of filtration). Particles will cake, increasing in thickness over time. At the same time, pressure drops increase and, after a certain time, they are such that it is no longer possible to continue filtration: this is called clogging. The filtration is then stopped to recover the solid. Before that, to purify the particle cake, one or more washing/spinning cycles are often carried out.

To facilitate the operation and increase the velocity of the fluid (which depends on the pressure drop in the porous medium), either a suction is applied downstream of the filter (vacuum filtration), or the pressure is increased upstream (pressure filtration). We can also use centrifugal force: this is a centrifugal filtration carried out in machines resembling bowl centrifuges, equipped with a filter cloth.

There are two modes of filtration through mechanical support:

– **constant-pressure filtration**: the pressure difference between the upstream and downstream sections of the filter is regulated to a constant value; as the thickness of the particle cake increases over time, the filtration rate (that is to say the production of filtrate) decreases under the effect of the increase in the pressure drop; it is the most used filtration mode in the industry;

– **constant-flow filtration**: the pressure difference between upstream and downstream sections of the filter **is increased over time** in order to maintain a constant flow despite the increase in pressure drop; this filtration method requires a somewhat complex regulation of pressure, according to the filtrate flow rate.

Industrially, there are also methods of filtration through mechanical support with variable flow and pressure.

There are many types of filter media: fibers (woven or felt), metal grids or perforated plates, paper, porous materials (for example, sintered), membranes.

FOCUS.– *Pre- and post-treatments.*

Drainage is often carried out before filtration in order to thicken (concentrate) the suspension and thus limit the flow rate to be treated. Flocculating agents can also be used in order to agglomerate small particles that are otherwise difficult to filter. If flocculation is impossible or too expensive, and if solids need not be recovered, adjuvants are used instead of using a very fine fabric to treat very fine particles. These high permeability compounds make it possible to delay clogging and thus facilitate filtration and filter cleaning.

In order to recover the filtrate or to free it of the particle cake, washing is carried out, that is to say the filtrate contained in the pores of the cake are eliminated by dilution. Centrifuging (or pressing) makes it possible to obtain the least moist particle cake possible, in order to reduce transport and/or drying costs. Finally, drying can be carried out if one wishes to recover a solid with very low residual moisture.

Many devices are available for carrying out filtration through mechanical support. The most common is the filter press (see Figure 5.27 in Box 5.3), which has a discontinuous operation, although it can be automated. It has the advantage of making it possible to develop very large filtration surfaces and to work under pressure.

Continuously operating filters include: belt filter, drum filter (Figure 5.12), disk filter, rotating table filter. For gas filtration: cartridge filter (or candle filter), bag or envelope filter.

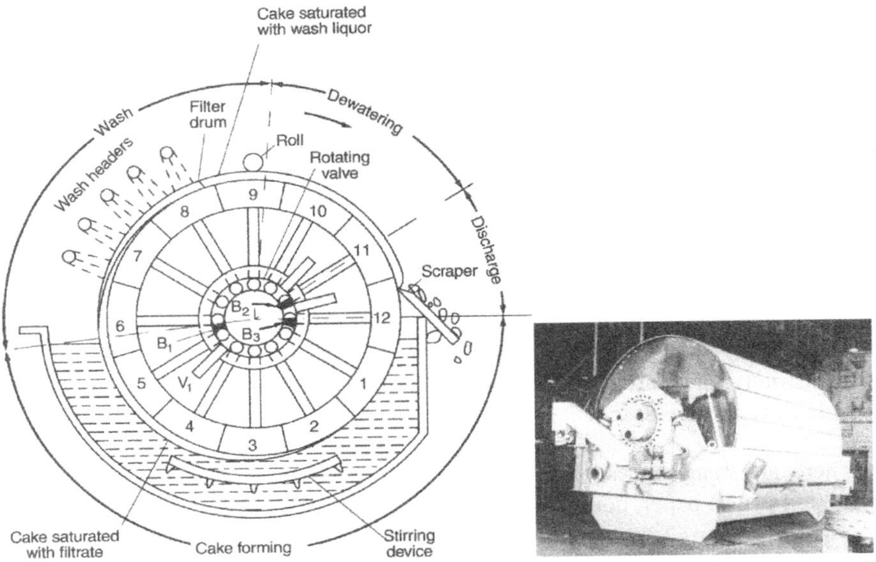

Figure 5.12. *Drum filter (source: Coulson and Richardson (2002))*

Videos
"Animation pour expliquer le fonctionnement des filtres-presse" – Génie des procédés, Cnam: https://youtu.be/2YNrOTZEooM "Horizontal Belt Filters" – FLSmidth: https://youtu.be/6voXE1HxYsY

In the **simplified theory of discontinuous filtration through a mechanical support**, the filtrate flow is by definition the volume of filtrate obtained per unit of time. It can also be expressed as the product of the filtration speed and the filtration area: $\frac{dV}{dt} = \Omega \cdot u$, where V is the volume of filtrate collected, t the time elapsed since the start of filtration, u the velocity of the liquid, and Ω the flow area. Thickness of the cake is denoted as Z.

It is assumed that filtration is carried out under the laminar regime (Stokes's law) in the cake and that the cake is homogeneous and incompressible.

Under these conditions, the pressure drop is proportional to the velocity and viscosity of the fluid (see Darcy's law in section 5.2.1): $\Delta P = R \cdot \mu \cdot u$. R is called resistance and denotes the proportionality factor (the unit of which in the international system is m^{-1}). This resistance can be broken down according to $R = R_s + R_g$, with R_s being the resistance of the medium and R_g that of the cake.

The specific resistance of the cake α in m kg^{-1} is such that $R_g = \alpha \cdot \frac{M}{\Omega}$, where M is the mass of dry cake (that is to say, counting only the solid). As the permeability of the cake B_0 in m^2 is such that $R_g = \frac{Z}{B_0}$, the expression of the specific resistance of the cake is $\alpha = \frac{1}{B_0 \cdot (1-\varepsilon) \cdot \rho_s}$, where ρ_s is the density of a solid particle (it therefore does not take account of the intergranular porosity).

With the mass of particles per unit volume of filtrate being denoted by $w = \frac{M}{V}$, the thickness of the cake is $Z = \frac{w \cdot V}{\Omega \cdot (1-\varepsilon) \cdot \rho_s}$.

With the ratio between mass of the wet cake and mass of the dry cake being denoted by m and the mass content of the solid material in suspension by s, we finally get $m = 1 + \frac{\varepsilon \cdot \rho_{liq}}{(1-\varepsilon) \cdot \rho_s}$ and $w = \frac{\rho_{liq} \cdot s}{1 - m \cdot s}$.

The differential equation of filtration is finally obtained:

$$\Delta P = \left(R_s + \frac{\alpha \cdot w}{\Omega} \cdot V \right) \cdot \frac{\mu}{\Omega} \cdot \frac{dV}{dt}$$

This differential equation can be integrated:

– in the case of constant-flow filtration:

$$\Delta P = R_s \cdot \mu \cdot u + \alpha \cdot w \cdot \mu \cdot u^2 \cdot t \text{, with constant } u = \frac{Q_v}{\Omega} = \text{constant}$$

– in the case of constant-pressure filtration:

$$\frac{t}{V} = \frac{R_s \cdot \mu}{\Omega \cdot \Delta P} + \frac{\alpha \cdot w \cdot \mu}{2 \cdot \Omega^2 \cdot \Delta P} \cdot V \text{, where } \Delta P \text{ is constant}$$

– for filtration with variable flow and pressure, the equation becomes:

$$\Delta P(t) = \frac{\mu}{\Omega} \cdot \left(R_s + R_g(t)\right) \cdot Q_v(t) \text{, with } R_g(t) = \frac{\alpha \cdot w}{\Omega} \cdot V(t)$$

It can no longer be integrated analytically and will require digital processing.

Generally, some filtration experiments are carried out in the laboratory on a sample of real suspension and with the same fabric as the future industrial operation, in order to determine the constant parameters of these equations (most often in the form of $\frac{R_s \cdot \mu}{\Omega}$ and $\frac{\alpha \cdot w \cdot \mu}{\Omega^2}$ groupings). We can then predict the behavior of the industrial filter for given operating conditions.

In the case of compressible cakes, one can generally find a power $0 < n < 0.8$ such that $\alpha \propto \left(\Delta P_g\right)^n$ and n is often about 0.1. We can then resume the filtration equation with α varying according to time.

5.2.5. Conclusion on fluid–solid mechanical separations

The few operations presented in this section are emblematic of the great diversity of mechanical operations using pulverulent solids. These operations are extremely common in material transformation plants, but, to this day, they remain much less scientifically evolved than heat and mass transfer operations involving only fluids, like distillation.

To learn more about fluid–solid mechanical separations

Coulson, J.M., Richardson, J.F. (2002). *Coulson & Richardson's Chemical Engineering*, 5th edition. Butterworth Heinemann, Oxford.

Fauduet, H. (2011). *Mécanique des fluides et des solides appliquée à la chimie.* Tec&Doc-Lavoisier, Paris.

Gatumel, C. *et al.* (n.d.). Self-training module: "Sciences et Technologie des Poudres" [Online]. École des mines d'Albi. Available at: https://nte.mines-albi.fr/STP/co/STP.html.

5.3. Stirring

In chemical engineering, fluid mechanics is involved at the level of connection equipment (piping, pumps, etc.); this concerns hydraulics, with the calculation of pressure drops and pump powers. But fluid mechanics also plays a very large role in all other equipment (reactors, columns, etc.), in particular by its influence on material and heat transfers; the study of flows in these devices is called hydrodynamics. In this regard, Professor Midoux wrote in the introduction to his book (Midoux 1985): "Fluid mechanics is a dry science for chemical engineers in the sense that it must be part of basic knowledge but, alone, does not solve 'noble' problems like the designs of a reactor or an exchanger. We cannot stress enough the fact that it is useless to design a device on paper by writing material and heat balances, while forgetting the unfortunate amount of movement that determines any convective transfer."

Stirring is a very old operation, which has gradually gone from being an art to a science, thanks to numerous experimental studies that were undertaken and the development of digital fluid mechanics tools. The operation consists of injecting, into a tank, the amount of movement caused by the dissipation of mechanical energy via a rotary part immersed in the medium. An agitation system will therefore consist of a tank and a shaft, equipped with an impeller, rotated by a motor.

Mixing in general and stirring, in particular, are very widespread in chemical engineering and can have various applications:

– mixing of miscible fluids (or homogenization): the quality of the mixture can be expressed in terms of micromixing scale. A mixing time constraint may also arise. Examples: neutralization, acidification, polymerization;

– suspension of pulverulent solids: this involves maintaining particles in partial or total suspension. It will be necessary to pay attention to the shear stress, in order to avoid damaging particles. Examples: crystallization, heterogeneous catalysis;

– liquid-liquid dispersion: in the case of immiscible liquids, one may wish to obtain a coarse dispersion (drop diameters > 10 μm, for example, for manufacturing glues, cosmetics, food products);

– gas-liquid dispersion: this involves bringing the gas and the liquid into contact, in order to achieve absorption in the best conditions, with or without a chemical reaction. Care should be taken not to clog the impeller. Examples: fermentation, oxidation.

For all these categories, once the primary objective has been met, we can focus on the issue of circulation, which will determine the material and heat transfers.

Industrially, for each of the main application categories, two types of problem can arise: design of a tank for a new application or scale-up of an existing installation.

Videos
"L'agitation en 15 min" – Génie des procédés, Cnam: https://youtu.be/y_IK-AVOa4s

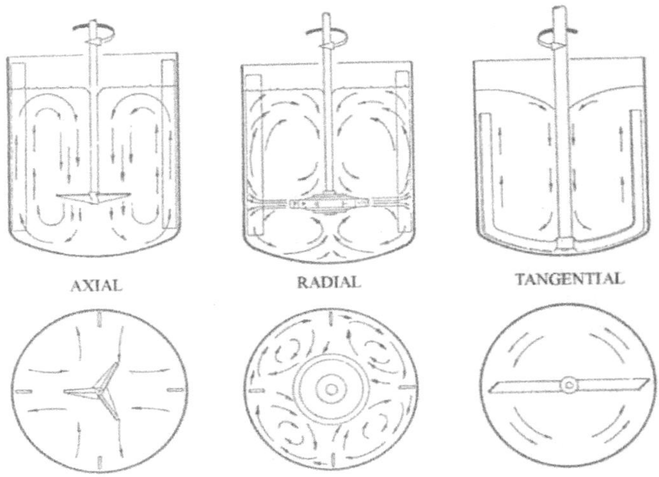

Figure 5.13. *Three main types of flow during agitation (source: Midoux (1996))*

5.3.1. *Qualitative aspects of stirring*

Three main types of flow are distinguished and illustrated in Figure 5.13:

– **axial**: the main flow develops parallel to the stirring shaft, generating a large circulation of liquid with a single circulation loop. This type of movement is obtained using an impeller from the propeller family (see Table 5.2);

– **radial**: the main flow develops perpendicular to the stirring shaft, generating a significant shear of the liquid around the impeller, with two circulation loops, one below and one above the impeller. This type of movement is obtained using an impeller from the turbine family, the best known being the Rushton turbine (see Table 5.2);

– **tangential**: the main flow develops tangentially, according to the rotation of the impeller, with good fluid renewal in the wall but limited mixing efficiency. This type of movement is obtained using scraping impellers like those shown in Figure 5.14; it is reserved for agitating very viscous fluids.

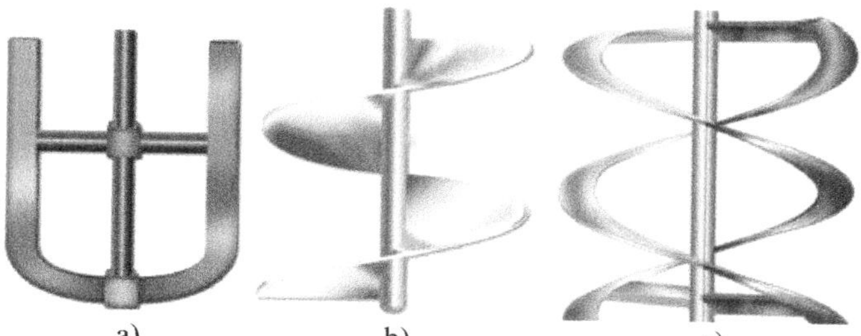

Figure 5.14. *Tangential impellers (Chemineer): (a) anchor; (b) screw; (c) double ribbon*

The **vortex** is a natural phenomenon in agitated tanks: under the effect of the impeller's rotation, the free surface of the liquid becomes hollow at the level of the stirring shaft, and the liquid rotates as a whole and that harms the mixing efficiency. It can be accompanied by absorption of the gas phase, which is not always desirable (see Figure 5.28 in Box 5.3). This phenomenon increases with the rotational speed. To combat vortex formation, it must be hindered by mechanical means: by tilting or making the impeller off-center, by placing a cross brace in the bottom of the tank, or by installing baffles (vertical blades) on the periphery of the tank, perpendicular to the wall (see first and second columns of Figure 5.13). Industrially, baffles are by far the most used to impede vortex formation, as they are extremely effective and also promote homogenization in the case of particle suspension, as illustrated by the following two videos.

Videos
"Baffles Demonstration 0 Baffle" – Jerwin van Dongen: https://youtu.be/ 0FF3Op2aWao "Baffles Demonstration 4 Baffles" – Jerwin van Dongen: https://youtu.be/l2D8SstFGtk

5.3.2. Quantitative aspects of stirring

FOCUS.– *Basic stirred tank.*

The basic stirred tank has well-established proportions (Figure 5.15):

– the height of the liquid z_L should be nearly equal to the diameter of the tank d_T (in the case of tall tanks, several impellers will be placed along the stirring shaft);

– the impeller must have a diameter $d_A = d_T / 3$ equal to about one third of that of the tank (in the case of tangential impellers, a much larger diameter will be chosen for the impeller, close to that of the tank);

– the impeller must be positioned at a height $z_A = z_L / 3$ equal to about one third of the filling of the tank;

– the baffles must have a lateral dimension b about a tenth of the diameter of the tank.

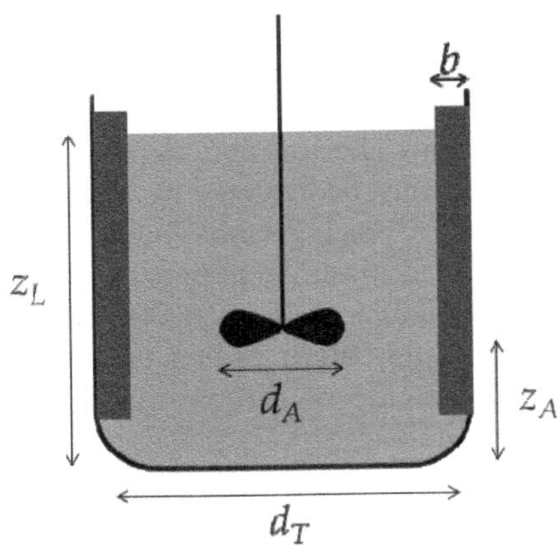

Figure 5.15. *Basic stirred tank*

The characteristic numbers of stirring such as the power number and the pumping number make it possible to select the impeller suitable for a given application.

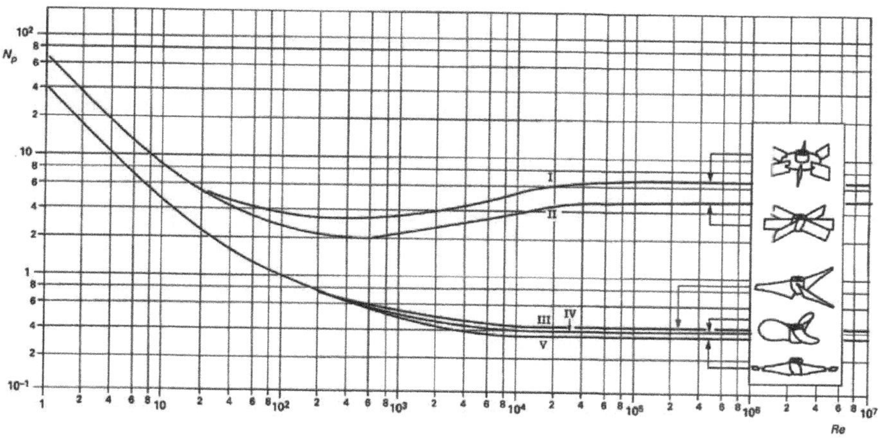

Figure 5.16. *Power number of a few impellers depending on the Reynolds number: I: Rushton-like disc turbine with 6 straight blades; II: 6-tilted blade turbine; III: propeller with 3 thin curved blades (large diameter); IV: 3-blade marine propeller; V: double-flow propeller with 2 blades (large diameter) (source: (Roustan 2005))*

The power curve plots the **power number** according to the Reynolds number:

– the Reynolds number in agitation is defined $Re = \frac{\rho \cdot N \cdot d_A^2}{\mu}$, where density ρ and viscosity μ are those of the agitated liquid or suspension, N is the rotational "speed" (it is a misnomer to call it "speed" because it is actually a frequency, expressed in s^{-1}), and d_A the diameter of the impeller;

– the power number is $N_p = \frac{P}{\rho \cdot N^3 \cdot d_A^5}$, where P is the agitation power or dissipated power; N_p can be seen as a drag coefficient of the impeller.

We observe in Figure 5.16:

– in the laminar regime ($Re < 10$), a decreasing linear dependence on a log-log scale, such that $N_p \propto \frac{1}{Re}$;

– in the turbulent regime ($Re > 10^4$), a plateau with constant $N_p = N_{p0}$: about 0.2 for an anchor; 0.3 to 1 for a propeller; 4 to 5 for a Rushton turbine.

The **pumping number** $N_{QP} = \dfrac{Q_p}{N \cdot d_A^3}$ is another important criterion in choosing an impeller. It is defined on the basis of the pumping flow Q_p: it is the liquid flow that actually passes through the impeller.

Table 5.2 presents the power and pumping numbers for some common impellers.

Illustration	**Impeller**	N_{p0}	N_{QP}
	Three-blade marine propeller	0.43	0.55
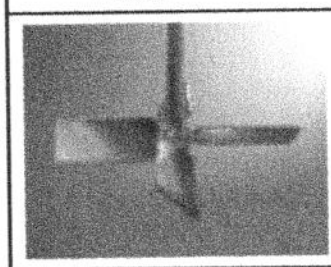	Propeller with four inclined blades	1.50	0.75
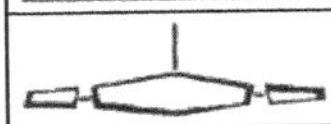	Two-blade double-flow propeller	0.32	0.58
	Propeller with three airfoil blades	0.40	0.65
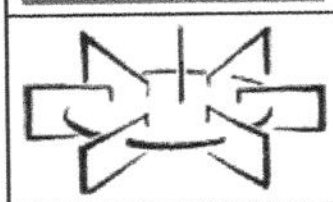	Six-blade Rushton turbine	5.0	0.68
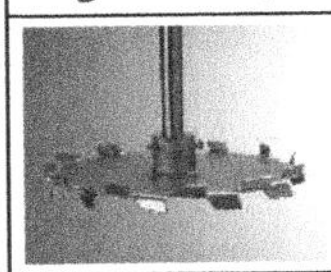	Grinding impeller	0.61	–
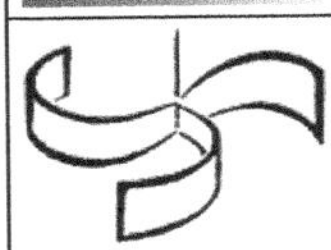	Curved blade turbine	2.1	0.50

Table 5.2. *Power and pumping numbers of various impellers in the turbulent regime (Missenard drawings; Lightnin photographs)*

5.3.3. *Choice of impellers*

Some very general (and therefore approximate) criteria can be given for choosing an impeller, according to the application:

– **homogenization**: during a homogenization operation in the turbulent regime, we seek to obtain a very good pumping rate, low shear, and low power. Suitable impellers will then be marine propellers and propellers with airfoil or inclined (large diameter) blades. For the homogenization of a viscous fluid (viscosity greater than 10 Pa s), good circulation and average shear will be sought, and consequently an average power, as the screw impeller and the helical ribbon impeller (see Figure 5.14) are then better suited. Finally, for homogenization with chemical reactions, correct circulation and high shear will be necessary, that is significant power; it will then be the Rushton turbine and propellers with airfoil or inclined blades that will be best suited;

– **suspension of particles**: to carry out operations in solid suspension, very good pumping and moderate shearing are required, therefore, average power. Airfoil, double-flow, or inclined blade propellers will give the best results. When the material transfer is limiting, it may be necessary to increase power and switch to turbines. Finally, to incorporate powder, a strong shear will be necessary; radial impellers, cones, and rotor turbines make it possible to carry out these operations;

– **fluid dispersion**: for gas-liquid dispersion or temporary liquid-liquid dispersion, high shear and correct circulation are required, therefore significant power. Turbines and cones will be the most suitable. For stable liquid-liquid dispersions, very high shear will be necessary; this will be obtained with small impellers such as cones, rotor turbines, or grinding impellers. In the case of small tanks for gas-liquid dispersion, airfoil or double-flow propellers or turbines with straight blades will be satisfactory.

5.3.4. *Stirred tank scale-up*

The extrapolation of stirring is not a problem that has been fully resolved; it largely depends on the application. However, two extrapolation methods are regularly used:

1) The conservation of the power per unit of volume: makes it possible to preserve a roughly uniform mixture and for which $N_{p1} \cdot N_1^3 \cdot d_{A1}^2 = N_{p2} \cdot N_2^3 \cdot d_{A2}^2$ (indices 1 and 2 correspond to the two size scales considered). This equation is simplified by keeping the agitation speed under the laminar regime and keeping the product $N^3 \cdot d_A^2$ under the turbulent regime.

2) The conservation of speed at the blade tip: $N_1 \cdot d_{A1} = N_2 \cdot d_{A2}$, which more or less allows for the conservation of the pump efficiency.

In both cases, the proportions of the tank will be conserved (geometric similarity).

5.3.5. *Conclusion on stirring*

Stirring, and more generally agitation, is one example of an operation dealing with contact phases. In the same category, we can find fluidization, packed or tray columns and many others. These operations will always be associated with other unit operations: distillation in columns, as previously mentioned, mixing, reaction or crystallization/precipitation in a stirred tank, drying or a heterogeneous reaction in a fluidized bed, etc.

To learn more about agitation

Xuereb, C., Poux, M., Bertrand, J. (2006). *Agitation et mélange. Aspects fondamentaux et applications industrielles*. Dunod-L'Usine Nouvelle, Paris.

5.4. Heat exchangers

A heat exchanger is a device that transfers heat between two fluids through a wall. Heat exchangers are present in most industrial processes. In fact, they can have different functions:

– preheating of a reaction medium;

– evaporating a liquid or condensation of a gas;

– cooling of reaction products;

– recovering energy;

– etc.

5.4.1. *Heat exchanger technologies*

There are many types of heat exchangers. Choosing the technology and the dimensions of the set-up depend on the problem to be addressed.

Videos
"Les échangeurs thermiques en 15 min" – Génie des procédés, Cnam: https://youtu.be/DqRYHwEAvdk

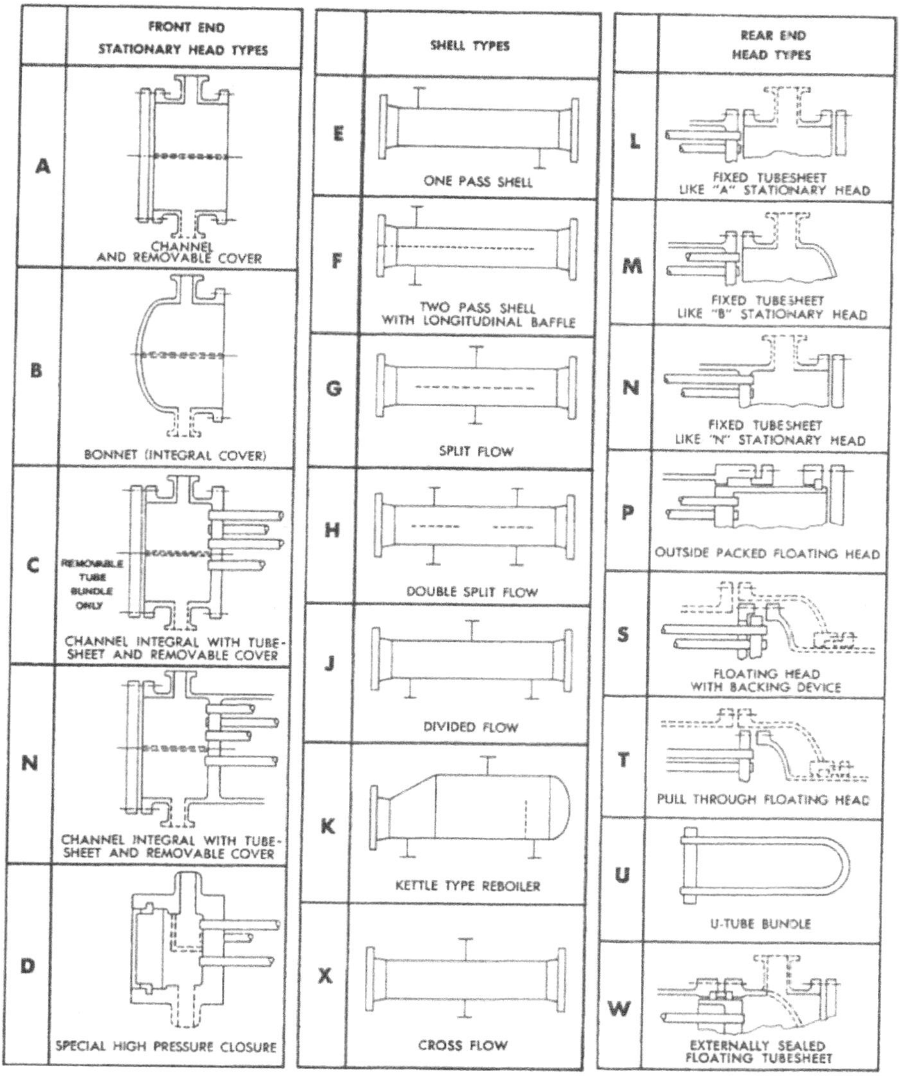

Figure 5.17. *TEMA standard for shell and tube heat exchangers (source: Perry and Green (1997))*

There are two main categories of heat exchanger technologies:

– for historical and economic reasons, **tubular exchangers** are the most widespread in the industry. The simplest tubular exchanger is the co-axial exchanger; it consists of two concentric tubes in which the two fluids circulate, in a co- or counter-current manner. Tube and shell heat exchangers are currently the most widespread; TEMA (Tubular Exchanger Manufacturers Association) has classified standardized shapes (Figure 5.17); this classification takes into account front end types (left column), shell types (central column), and rear end types (right column). Some tubular exchangers use the principle of fins to promote the efficiency of heat transfer;

– **plate heat exchangers** were developed more recently. They have the advantage of being more compact and offering significantly higher heat transfer coefficients than tube and shell heat exchangers. However, unless the plates are welded or brazed (which poses a problem in the event of fouling), they cannot withstand high pressures and are not suitable for evaporation. The maximum operating temperature is limited by the nature of the gaskets.

Videos
"Fonctionnement d'un échangeur à plaques" – Génie des procédés, Cnam: https://youtu.be/1UWcJ-Qxgn8

5.4.2. *Designing heat exchangers*

Heat exchanger calculations are based on heat transfer concepts mentioned in section 4.2.2.

In the absence of heat loss to the outside environment, the energy balance for a heat exchanger is written as:

$$\phi = Q_{m,c} \cdot Cp_c \cdot \left(T_c^e - T_c^s\right) = Q_{m,f} \cdot Cp_f \cdot \left(T_f^s - T_f^e\right)$$

where the temperature index designates the hot fluid (c) or the cold fluid (f) and exposes it to the input (e) or the output (s); Q_m is mass flow and C_p specific heat capacity.

FOCUS.– *Typical temperature profiles.*

Typical temperature profiles are shown diagrammatically in Figure 5.18.

The advantage of the counter-current exchanger is, firstly, that it allows for the temperature of the hot fluid at the output to be much lower than the temperature of the cold fluid at the output, and, secondly, that it maintains a significant temperature difference between the hot fluid and the cold fluid throughout the exchanger: however, it is this difference in temperature that is the driving force behind the transfer.

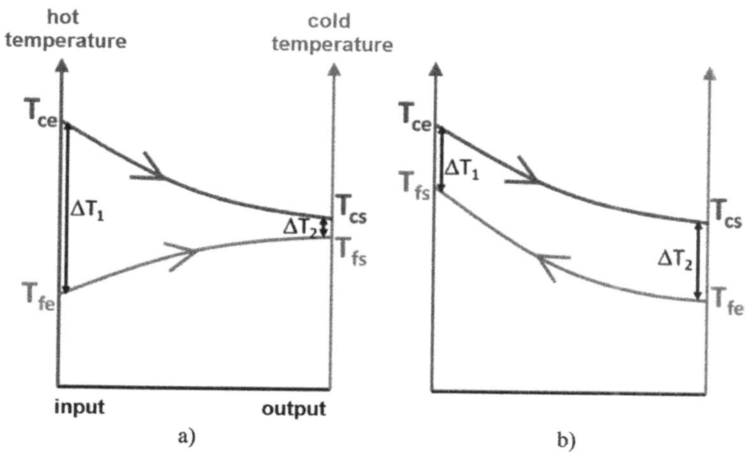

Figure 5.18. *Typical temperature profiles in heat exchangers: (a) co-current; (b) counter-current. For a color version of this figure, see www.iste.co.uk/dalpont/process1.zip*

For a coaxial exchanger, we can show (see Box 5.4) that this transferred heat flow is also equal to $\phi = H \cdot S \cdot \Delta T_{ml}$, where:

– H is the overall heat transfer coefficient taking into account the convection on either side of the wall (in the hot fluid and in the cold fluid) and the conduction in the wall. There are numerous correlations in the literature enabling these transfer coefficients to be estimated according to the configuration of the exchanger, and the Reynolds number that characterizes the flow of the fluid;

– S is the transfer surface, which will be the quantity that we will seek to calculate during a design phase;

– $\Delta T_{ml} = \dfrac{\Delta T_1 - \Delta T_2}{ln\dfrac{\Delta T_1}{\Delta T_2}}$ is the logarithmic mean of the temperature difference between the hot fluid and the cold fluid, with ΔT_1 and ΔT_2 this difference in temperature on one side and the other of the exchanger.

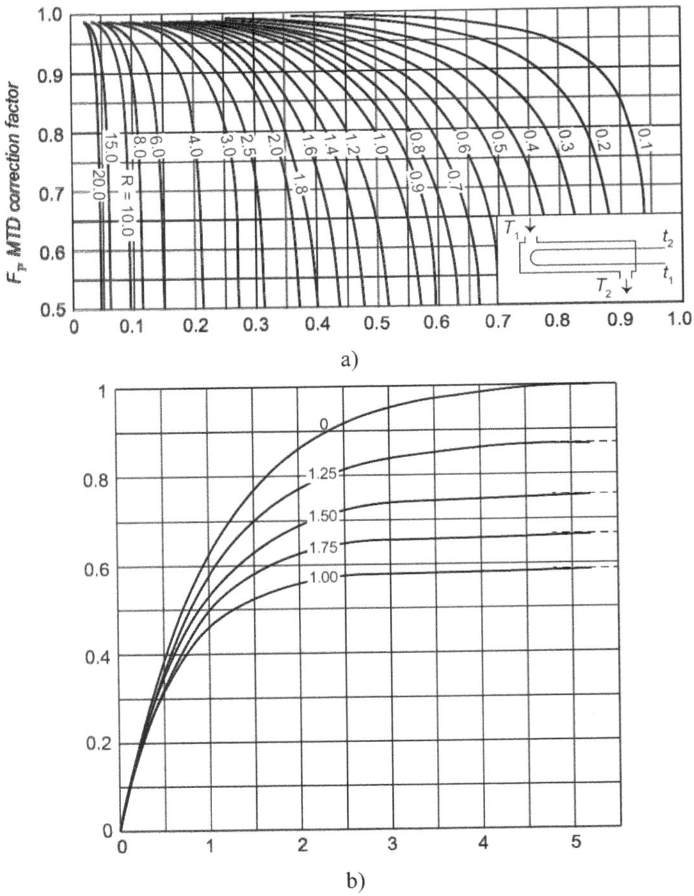

Figure 5.19. *Nomograms for the calculation of a multitubular exchanger 1 shell pass – 2 tube passes: a) correction factor method (Perry and Green 1997): eff in X axis and F in Y axis; b) eff-NUT method (Coulson and Richardon 2002): NUT in X axis; eff in Y axis*

For other types of exchangers, we can use:

– **the correction factor method** for which $\phi = F \cdot H \cdot S \cdot \Delta T_{ml}$, and the correction factor F is evaluated using nomograms like those in Figure 5.19a, where

$$eff = \frac{T_f^s - T_f^e}{T_c^e - T_f^e};$$

– **the eff-NUT method** for which $\phi = eff \cdot (Q_m \cdot Cp)_{min} \cdot \left(T_c^e - T_f^e\right)$, and the efficiency *eff* is evaluated using nomograms like those in Figure 5.19b, where $NUT = \dfrac{H \cdot S}{(Q_m \cdot Cp)_{min}}$.

The different curves on these nomograms correspond to different values of the heat ratio R, which can be calculated in the first case according to $R = \dfrac{T_c^e - T_c^s}{T_f^s - T_f^e}$ and in the second case, if the output temperatures are not known, according to $R = \dfrac{(Q_m \cdot Cp)_{min}}{(Q_m \cdot Cp)_{max}}$.

5.4.3. *Conclusion on heat exchangers*

Heat exchangers are one of the "ancillary" operations essential to the running of a manufacturing plant. Another example is the transport of fluids and powders.

To learn more about heat exchangers

Bontemps, A. *et al.* (2014). *Articles BE9515 à BE9519*. Techniques de l'Ingénieur, Saint-Denis.

Leleu, R. (2002). *Conception et technologie des systèmes thermiques*. Hermes-Lavoisier, Paris.

5.5. Reactors

Reactors are the equipment in which reactions are carried out. In this section, we will focus on chemical reactions; biological and biochemical reactions are discussed in Chapter 1 of Volume 2. Chemical reaction engineering is the branch of chemical engineering that deals with reactors. Since Levenspiel (1926–2017) (Levenspiel 1999), the study of reactors is by no means limited to a collection of "recipes", but endeavors to propose a "method", which Jacques Villermaux called "the royal way" in the epilogue of his book (Villermaux 1993).

A distinction is made between homogeneous chemical reaction engineering – CRE (single-phase reactors: liquid or gaseous) – and heterogeneous (chemical) reaction engineering – HRE (multi-phase reactors, the most common of which are solid catalyst reactors, consumable solid reactors, gas-liquid reactors).

One or more chemical reactions can take place in these reactors. These can be complete (they occur in one direction) or reversible reactions (likely to occur in both directions), exo- or endothermic reactions (athermic reactions are rare); these characteristics are linked to the thermodynamics of these chemical transformations. The reactor can operate in the transient or stationary regime; it can be closed, semi-closed, or semi-open, or even open (see Figure 4.18).

Inside a reactor, the mixing efficiency can be perfect (ideally stirred-tank reactor – ISTR or continuous stirred-tank reactor (CSTR) if the ideally stirred-tank reactor is open), or the mixing efficiency can be nil (plug flow reactor), or in between. The ISTR and plug flow reactor are called ideal reactors; they are used to describe the behavior of the two extreme cases in terms of mixing in the reactors. Real reactors can be described based on these ideal reactors, as we will see in section 5.5.3.

5.5.1. *Conversion rate and generalized extent of reaction*

The **conversion rate** X_A of any reactant A can be defined according to the following expressions:

– $n_A = n_{A0} \cdot (1 - X_A)$ in a closed reactor;

– $F_A = F_{A0} \cdot (1 - X_A)$ in an open reactor.

n_j denotes number of moles of the species j and F_j its molar flux. The index 0 corresponds to the reference state. The reference state chosen is most often the initial instant in the case of a discontinuous operation, and the feed to the reactor in the case of a continuous operation in the permanent regime. When the reactive fluid is gaseous, the reference state is generally taken under the standard conditions of temperature and pressure.

We can then express the number of moles or the molar flux of any species j relative to the conversion rate of species A, which then plays a particular role and will be called "key reactant":

– $n_j = n_{j0} - \dfrac{\nu_j}{\nu_A} \cdot n_{A0} \cdot X_A$ in a closed reactor;

– $F_j = F_{j0} - \dfrac{\nu_j}{\nu_A} \cdot F_{A0} \cdot X_A$ in an open reactor.

ν_j represents the stoichiometric coefficient (algebraic: positive for reaction products and negative for reactants) of the species j in the reaction carried out in the reactor. To simplify these, we can rewrite the reaction equation so that the stoichiometric coefficient of reactant A is equal to -1.

To study multi-reaction systems, we may prefer using the concept of **generalized extent** X_i of reaction i. The number of moles or the molar flux of any species j will then be written as:

– $n_j = n_{j0} + n_0 \cdot \sum_i \nu_{ij} \cdot X_i$ in a closed reactor;

– $F_j = F_{j0} + F_0 \cdot \sum_i \nu_{ij} \cdot X_i$ in an open reactor.

ν_{ij} represents the stoichiometric coefficient (always algebraic) of species j in reaction i. In the presence of inert species, that is to say of species that are not involved in any reaction, n_0 is the total number of moles of species present *excluding* inert species in the reference state; F_0 is the total incoming molar flux *excluding* inert species. The number of moles and the molar flux of inert species are denoted as n_I and F_I, respectively.

It can be shown that the total number of moles and the total molar flux (useful quantities in the gas phase to express the concentrations of species with reference to the ideal gas law) are then expressed as follows:

– $n_{total} = n_I + n_0 \cdot \left(1 + \sum_i \Delta\nu_i \cdot X_i\right)$ in a closed reactor;

– $F_{total} = F_I + F_0 \cdot \left(1 + \sum_i \Delta\nu_i \cdot X_i\right)$ in an open reactor.

$\Delta\nu_i = \sum_j \nu_{ij}$ is the stoichiometric dilation of reaction i, that is to say the algebraic sum of the stoichiometric coefficients of the various species in this reaction.

In the presence of inert species, that is to say species that are not involved in any reaction, n_0 is the total number of moles of species present *excluding* inert species (the term "active" is sometimes used) in the reference state; F_0 is the total incoming molar flow *excluding* inert species. The number of moles and the molar flux of inert species are denoted as n_I and F_I, respectively.

5.5.2. *Ideal homogeneous reactors*

To carry out homogeneous phase reactions, there is little technological diversity. Thus, the construction of a well mixed reactor in the gas phase can be limited to an enclosure into which these gases are injected at high velocity; systems like static mixers can possibly be used to facilitate mixing. For the construction of a well mixed reactor in the liquid phase, a stirred tank will be used (see section 5.3; see also Figure 5.23). Finally, tubular reactors, simple tubes with a large length to diameter ratio, will be used in the liquid or gas phase to create reactors similar to a plug flow; they may contain packing material.

Among the technologies under development, mention can be made of the plate reactor: the reaction is carried out in a plate exchanger (see section 5.4.1), in order to benefit from the qualities of excellent heat transfer in this type of contactor. We thus get a multifunctional and intensified reactor (see Box 5.1).

Videos
"Static Mixer" – Samhwamix: https://youtu.be/4H2Vk7_cCCc

A **material balance** consists of writing the conservation of mass (see Box 4.1). However, as soon as we have to work with a reactor, where one or more reactions take place, we need to think in terms of molar flux. The general balance is then written as:

$$F_j^e + \left(\sum_i \nu_{ij} \cdot r_i \right) \cdot V = F_j^s + \frac{dn_j}{dt}$$

where the exponents e and s denote the input and the output of the system, F_j and n_j are the molar flux and the number of moles, respectively, of species j, ν_{ij} is the stoichiometric coefficient of species j in reaction number i, r_i is the rate of reaction number i (see section 4.2.3), and finally V is the volume of the reaction medium.

It must be noted that it is only possible to write this balance for a uniform system. Depending on the type of reactor, this balance may therefore be written for the complete reactor (this is the case for an ideally stirred-tank reactor) or only on a portion of the reactor (small enough to be considered uniform). In a tubular plug flow reactor, this balance will be written for a reactor section.

Videos
"Les bilans en chimie" – Thierry Meyer, EPFL: https://youtu.be/JxLA1JZnSbg

The previous material balance in species j could be broken down according to the different types of ideal reactors, where a single reaction with rate r takes place as follows:

– $\nu_j \cdot r \cdot V = \dfrac{dn_j}{dt}$ for a closed ideally stirred tank reactor;

– $F_j^e + \nu_j \cdot r \cdot V = F_j^S$ for a continuous stirred tank reactor (CSTR) operating in the permanent regime;

– $\nu_j \cdot r = \dfrac{dF_j}{dV}$ on a section of a plug flow reactor in the permanent regime.

In the first and third cases, it will be necessary to go through an integration step (possibly numerical, depending on the complexity of the speed law used to explain r depending on the concentrations of different reactants, even reaction products): integration in time to take into account the variation in the composition of the reaction medium over time for the first case, and integration into space for the third, to take into account the variation in the composition of the reaction medium, according to the zones of the reactor. In the second case, no integration is necessary, as the continuous stirred tank reactor operates under the output conditions, that is to say that the same concentrations and the same temperature prevail everywhere in the reactor as at its output.

From this material balance, and its integration where appropriate, we can, for example:

– predict the conversion rate of any species (or the extent of a reaction) at the output of the reactor of a known size;

– design the reactor (that is to say calculate its volume) to achieve a desired conversion rate (or extent);

– obtain the value of the rate constant under the operating conditions of the reactor of known size, by measuring the conversion rate at the output.

In the case of a reactor in which only one reaction takes place, simply knowing the rate of reaction (see section 4.2.3) makes it possible to predict which type of ideal reactor will have the smallest volume for a given output conversion rate; for this purpose, it suffices to plot the "**Levenspiel plot**", C_{A0}/r vs X_A.

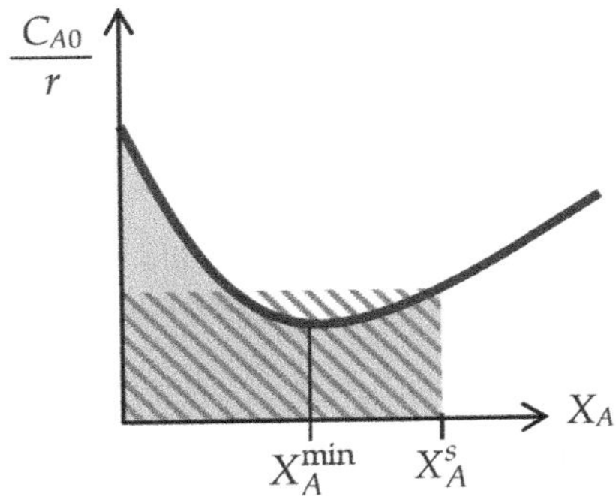

Figure 5.20. *"Levenspiel plot" the case of a curve with a minimum. For a color version of this figure, see www.iste.co.uk/dalpont/process1.zip*

When this curve is monotonically increasing, the plug flow reaction will always have the smallest volume to reach a given conversion rate. When the curve is monotonically decreasing, it will be the CSTR.

When the curve presents a minimum, the graph must be studied in more detail; the passage time (ratio between the reaction volume and the volume flow through it) is indeed equal to:

– the area under the curve (shaded yellow in Figure 5.20) in the case of a plug flow reactor;

– the area of the rectangle (shaded green in Figure 5.20) in the case of a CSTR.

We could even consider an optimal combination of ideal reactors to further minimize the reaction volume: a CSTR to convert up to X_A^{min}, then a plug flow reactor to complete the conversion up to X_A^s.

A reactor is often where several reactions take place. However, generally only one product is of interest; reactions that do not lead to it or degrade it are then called parasitic reactions. The reaction leading from the reactant(s) to the product of interest is called the main reaction. A reaction that consumes a useful reactant without producing the product of interest is said to be competing or parallel. A reaction that consumes the product of interest is called a consecutive or serial reaction.

In the presence of parasitic reactions, the concept of rate of conversion or extent is no longer sufficient to measure the efficiency of the reactor for manufacturing the product of interest. We must then define other quantities. In open reactors, the **overall efficiency** (also called the useful product reaction rate) is defined as:

$$Y_{P/A} = \frac{F_P}{\nu_{P/A} \cdot F_{A0}}$$

where P is the desired product, A the key reactant, and $\nu_{P/A}$ the maximum number of moles of P that can be obtained from one mole of A. $Y_{P/A}$ is always less than 1, because other reactions can consume A, and/or not all of A is converted.

We can also define a **relative local efficiency**, which is none other than the derivative of the overall efficiency:

$$\eta'_{P/A} = \frac{R_P}{\nu_{P/A} \cdot (-R_A)}$$

where $R_j = \sum_i \nu_{ij} \cdot r_i$ the overall production speed (in the algebraic sense) of species j.

In the case of a closed reactor, it suffices to replace, in all these definitions, the molar fluxes F by numbers of moles n.

We can therefore calculate the overall efficiency by integrating the relative local efficiency, which depends on the speed laws of various reactions taking place: $Y_{P/A} = \int_{X_A^e}^{X_A^s} \frac{d\eta'_{P/A}}{dX_A}$. As with the "Levenspiel plot", a graphical representation can be used to optimize the efficiency of a reactor or a combination of reactors, with the difference that it is a question here of maximizing the efficiency (whereas the passage time or the volume had to be minimized with the "Levenspiel plot"). In CSTR, as all the compositions are uniform, the overall efficiency corresponds to the area of the rectangle shaded green, and in a plug flow reactor, in the area shaded yellow (see Figure 5.21). When the relative local efficiency is decreasing, the plug flow reactor would be more useful; when it is decreasing, it would be the CSTR. When this curve presents a maximum as in Figure 5.21, the maximum overall efficiency will be obtained by first using a CSTR to convert up to $X_A^{\max}$ and then using a plug flow up to X_A^s.

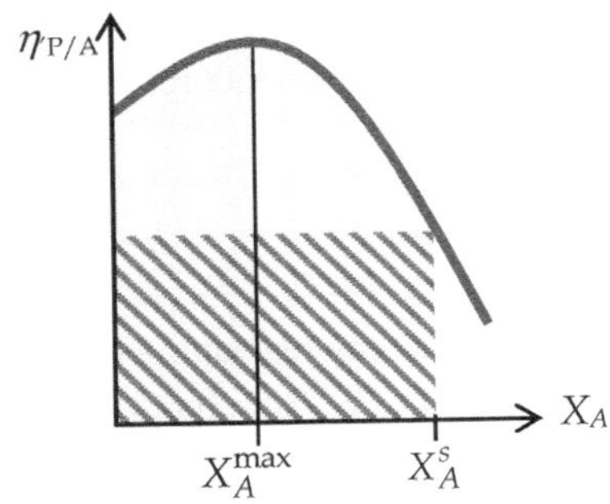

Figure 5.21. *Calculation and optimization of the overall efficiency from the relative local efficiency. For a color version of this figure, see www.iste.co.uk/dalpont/process1.zip*

Unfortunately, the reactor or the optimum combination of reactors, in terms efficiency, rarely coincides with the minimum volume. It is therefore necessary to make an economic optimization, taking into account the investment and operating costs of various equipment in the plant, for both reactors and the separators that follow it. We thus understand the importance of properly designing the reactor, because even if it does not generally represent the capital expenditure of a new manufacturing plant, it is possible to greatly minimize downstream separation if the reactor is well designed.

The first principle of thermodynamics allows, by means of a few approximations (negligible variations of potential energy and kinetic energy, absence of change of state and heat capacities considered constant in the temperature domain studied) and work on the expression of enthalpy, state function, to arrive at writing the **heat balance** for an open reactor operating in the permanent regime:

$$\varphi = Q_m^e \cdot \mathrm{Cp} \cdot \left(T^S - T^e\right) + F_0 \cdot \sum_i \Delta \mathrm{r} H_i(T^S) \cdot X_i$$

where φ the heat flux from the exterior, Q_m^e is the mass flow entering the reactor and Cp the average specific heat capacity (per unit of mass) at the input to the reactor, T^e and T^s are the temperatures of the reaction medium at the input and the output of the reactor, F_0 is the molar flux of active ingredients entering the reactor (excluding inert ingredients) and X_i the extent of each of the reactions number i, and $\Delta \mathrm{r} H_i(T^S)$ is the molar enthalpy of the reaction number i to the output temperature of the reactor.

An important question arises about the **thermal operation of reactors**; this concerns the choice of working temperature:

– for reactors with a main endothermic or irreversible exothermic reaction, it is necessary to work isothermally, at the maximum temperature that can be safely tolerated by the chemical species used, and by the reactor. Indeed, the temperature is then favorable both from a thermodynamic and kinetic point of view;

– for reactors with a main reversible exothermic reaction, conversion must be started at high temperature in order to take advantage of the highest reaction rates, then the working temperature must be gradually decreased in correlation to the thermodynamic limit (the temperature is unfavorable from a thermodynamic point of view). There is an "optimal thermal path" called optimum temperature progression (OTP) for this kind of reaction. It is however very difficult to follow this OTP precisely, which is why, during reversible exothermic reactions, a cascade of a few adiabatic reactors (often 3 or 4) will often be used with intermediate coolers.

5.5.3. *Non-ideal reactors*

Measuring the residence time distribution (RTD) is a simple and effective method for studying macroscopic flows in real reactors (see Figure 5.25 in Box 5.3). It allows for the detection of faults in the flow (short-circuit, recirculation, dead zones) and hydrodynamic modeling. This method, developed on fluids, could be extended to the study of the flow of granular solids in processes. Initially invented for studying reactors, this systemic method for characterizing macroscopic flows has seen its use extended to all equipment in the process industries as well as studies in the natural environment.

The principle of the measurement is to "mark" molecules that enter the system (injection, pulse, or step) and count them in the output flow according to the time. The tracer must have the same hydrodynamic properties as the "fluid" studied; it must be detectable by a physical property, stable, and perfectly neutral for the system. The conditions of application of the method are a flow in a steady and deterministic regime, an incompressible fluid, a flow without diffusion at the input and the output, and detection in a well-mixed area. A more sophisticated version of the method makes it possible to adapt it to compressible fluids.

The distribution function of the residence times E is such that the fraction of fluid that remains in the equipment for a time between t_s and $t_s + dt_s$ is equal to the product $E(t_s).dt_s$. In the case of a tracer injection in the form of a pulse, it is easily calculated from the tracer C concentration measured at the output of the equipment:

$$E\,(t_S) = \frac{C\,(t_S)}{\int_0^\infty C(t)\cdot \mathrm{d}t}$$

When the injection is in the form of a step, the analysis of the tracer concentration at the output of the equipment makes it possible to return to function F, which is none other than the integral of the distribution function:

$$F(t_S) = \int_0^{t_s} E(t) \cdot \mathrm{d}t$$

When the input of the system is not accessible to perform an injection, we can use the two-point method (concentrations of tracer C^e at the input of the equipment and C^s at the output) and go back to function E by deconvolution, since the C^e and C^s concentrations are linked by the convolution product:

$$C^S(t_S) = \int_0^{t_s} C^e(t) \cdot E(t_S - t) \cdot \mathrm{d}t$$

In Figure 5.22, RTD forms for the two ideal reactors are represented: plug flow reactor in blue (the Dirac function δ is simply shifted in time) and CSTR in green (the RTD function is exponentially decreasing). A typical shape for any reactor is also represented in red in this figure: a peak more or less around the average residence time $\overline{t_S}$ and a more or less significant "drag" at large t_S.

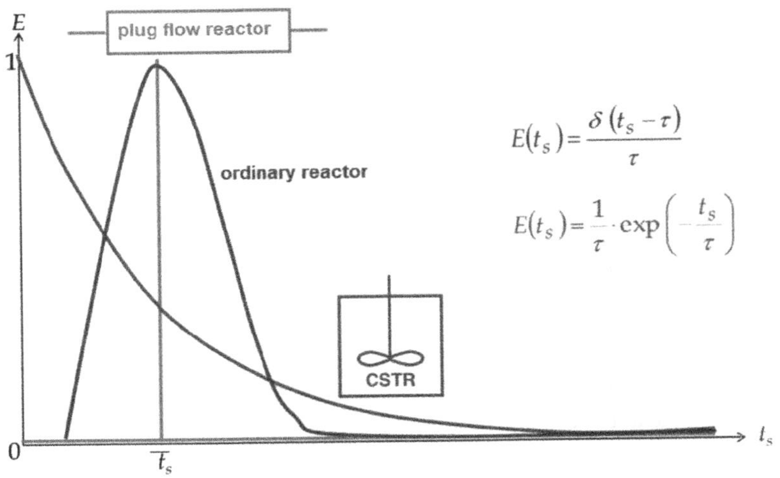

Figure 5.22. *Typical residence time distributions. For a color version of this figure, see www.iste.co.uk/dalpont/process1.zip*

In the absence of a short circuit and recirculation or dead volume, the mean residence time $\overline{t_s}$ is equal to the time of passage τ (which is itself equal to the ratio between the useful volume of the equipment and the volume flow passing through it). A mean residence time less than the time of passage is typical in the presence of recirculations and/or dead volumes; a mean residence time greater than the time of passage is indicative of short circuits.

The quantitative analysis of the RTD curve of a real reactor makes it possible to model it using a set of elementary patterns (ideal reactors: plug flow reactor and CSTR, as well as the dispersive plug flow reactor) and associations (in series, parallel; with exchange, short-circuit, or recirculation).

This is why the study of homogeneous reactors is limited to the study of ideal reactors: we can reconstruct the real reactor by assembling ideal reactors.

Videos
"5 min pour comprendre : le réacteur piston" – Génie des procédés, Cnam: https://youtu.be/7bM_nC3YPiU "La DTS en 13 min" – Génie des procédés, Cnam: https://youtu.be/1faiuMWgel8

5.5.4. *Multi-phase reactors*

In heterogeneous chemical reaction engineering, there is an additional problem to chemical reactions in terms of access of reactants from one phase to those of another phase; therefore, physical kinetics are added to chemical kinetics (convection and diffusion of matter, as well as heat; see section 4.2.2).

Referring back to heterogeneous catalysis, (section 4.2.4 and in particular Figure 4.14) let us take the case of a reaction involving a solid catalyst.

There are usually seven stages for carrying out a catalytic reaction.

We start by studying the phenomena on the scale of a catalyst pellet:

1) The reactive species or species present in the fluid phase come(s) to the surface of the catalyst pellet (external transfer phenomenon).

2) The reactive species, or species present in the fluid phase, migrate(s) inside a pore of the catalyst pellet by diffusion.

3) The reactive species, or species present in the fluid phase, adsorbs(s) on an active catalyst site.

4) The reaction takes place.

5) The non-solid reaction products desorb.

6) The reaction products migrate towards the outside of the catalyst via a pore of the catalyst pellet by diffusion.

7) The reaction products move away from the catalyst pellet (external transfer phenomenon).

Steps 3 to 5 are often combined under an **apparent reaction rate** $\bar{r}$ which, over a limited concentration range, can be reduced to about *n* order with respect to the reactant fluid *A*: thus $\bar{r} = k \cdot C_A{}^n$, where *k* generally no longer follows the Arrhenius law for temperature dependence.

When we model the diffusion in the catalyst pellet (steps 2 and 6) and the external transfer around the grains (steps 1 and 7), we are led to define a dimensionless number called Thiele modulus, as well as internal η_s and external η_e, **efficiency factors**, which link the rates r_s under the surface conditions of the grain (concentration C_s and temperature T_s) and r_e far from the grain (concentration C_e and temperature T_e) at the apparent rate $\bar{r}$:

$$\bar{r} = \eta_e \cdot r_e = \eta_s \cdot r_s \; ; \eta_e = \frac{k_D}{k_D + k \cdot L}$$

where k_D is the external material transfer conductance, *k* the rate constant defined above, and *L* the characteristic dimension of the catalyst pellet (for example, its diameter if it is spherical).

The **Thiele modulus** is:

$$\varphi_s = \left(\frac{r_s L^2}{D_e C_s}\right)^{1/2}$$

where D_e is the internal diffusion coefficient.

As mentioned in section 4.2.4, the catalyst pellet can have various shapes. According to this shape, the expression of the internal efficiency factor according to the Thiele modulus varies:

– $\eta_S = \frac{3\phi_S \coth(3\phi_S)-1}{3\phi_S^2}$ for a spherical catalyst pellet;

– $\eta_S = \frac{th\phi_S}{\phi_S}$ for wafer-shaped catalyst with a thickness of 2.L.

When $\phi_s^2 << 1$, $\eta_s = 1$, it is the **chemical regime**: the concentration of fluid reactant is the same in the pores of the catalyst and far from the catalyst pellets, and rate of the process is decided by chemical kinetics.

When $\phi_s^2 >> 1$, $\eta_S = \frac{1}{\phi_S}$, it is **diffusion-controlled** (or "physical" regime): the reaction is fast with respect to the diffusion, which imparts its rate to the overall process. We generally consider the **mixed regime** for $0.3 < \phi_s < 3$.

The Prater thermicity criterion $\beta_s = \frac{D_e C_S(-\Delta rH)}{\lambda_e T_S}$ (where ΔrH is the enthalpy of reaction and λ_e the thermal conductivity of the catalyst) makes it possible to evaluate the maximum temperature gradient in a catalyst pellet.

Solid catalyst reactions can be carried out in fixed, fluidized, or transported bed reactors, or one that is structured (connecting fixed bed reactors with internal and/or external heat exchangers). Various criteria make it possible to select the most suitable reactor according to the intended application, like the technology that allows for significant heat exchange, or the overall operating cost (often linked to the pressure drop in the device, catalyst particle recovery, or regular replacement of the catalyst, due to its deactivation):

– when the reaction rate is low, it is very likely that it will control the overall rate, and thus it is useful to try to promote transfers; large particles are preferable because they generate less pressure drop and the fixed bed reactor is the simplest;

– when the catalyst deactivates quickly, it must be possible to regenerate it often: a moving bed will be preferable, but it must be able to separate the catalyst and the fluid;

– when there is no selectivity problem, a stirred tank reactor will be preferable. A reactor similar to the plug flow reactor will be chosen, if the desired product is an intermediate reaction;

– when the internal transfer is slow, it is advantageous to use shell impregnated (and not mass) catalysts;

– when there is a risk of runaway, provision must be made for cooling in all parts of the equipment (multitubular bundle); otherwise, for a simple exothermic reaction, intermediate exchangers may suffice;

– the fluidized bed reactor allows for better temperature control than the fixed bed reactor, but rather it is a stirred-tank kind of reactor, so it is not well suited in the event of selectivity or runaway problems.

The design of catalytic reactors is complex and will not be detailed here. Interested readers can refer, in particular, to the self-training module referenced below. This module also details the case of consumable solid and gas-liquid reactors.

To learn more about reactors

Belandria, V., Billet, A.-M., Debacq, M., Le Coq, O., Schaer, E. (2019). Self-training module "GRCpoly : les réacteurs polyphasiques" [Online]. Available at: https://sites.cnam.fr/industries-de-procedes/co/GRCpoly.html.

Levenspiel, O. (1999). *Chemical Reaction Engineering*. Wiley, New York.

Villermaux, J. (1993). *Génie de la réaction chimique*. Tec&Doc-Lavoisier, Paris.

5.6. Conclusion

The purpose of this chapter was to give an overview (by no means exhaustive) of the unit operations encountered in chemical engineering. The examples chosen made it possible to show operations of various kinds, as well as at different stages of scientific evolution. Readers interested in these operations can refer to the works cited in each of the "To learn more about" sections. To discover other unit operations, some books are listed below.

To learn more about unit operations in chemical engineering

Collectif (2004). *Ullmann's Processes and Process Engineering*, 1, 2 and 3. Wiley-VCH, Weinheim.

Coulson, J.M., Richardson, J.F. (2002). *Coulson & Richardson's Chemical Engineering*, 5th edition. Butterworth Heinemann, Oxford.

Green, D.W., Southard, M.Z. (2019). *PERRY'S Chemical Engineers' Handbook*, 9th edition. McGraw-Hill, New York.

McCabe, W.L., Smith, J.C. (2001). *Unit Operations of Chemical Engineering*, 6th edition. McGraw-Hill, New York.

Ronze, D. (2008). *Introduction au génie des procédés*. Lavoisier, Paris.

5.7. Boxes

Process intensification

The first books on process intensification date back to the late 1970s, when the company ICI showed that this concept was an effective way to reduce capital costs of a process by developing HiGee technology, to separate mixtures by centrifugation. The large-scale application of this technology has made it possible to replace 30-meter separation columns with rotary devices with a diameter of 1.5 meters for the same efficiency.

The foundations of process intensification were theorized in the 1980s by Ramshaw (Ramshaw 1993; Jachuck and Ramshaw 1994) and then by Stankiewicz and Moulijn (2000), who proposed a definition, the specificity of which lies in the "significant reduction in the size/capacity ratio". Whether this ratio is reduced by increasing the capacity of equipment, keeping the size a constant, or by reducing its size keeping the capacity a constant, there appear two central concepts in intensification, which are the miniaturization of equipment and the acceleration of phenomena. This definition was then extended to processes thanks to multi-scale approaches and the search for synergies between functional units. The intensification of processes therefore consists, through the development of suitable methods and devices, of designing more efficient, more compact, and more economical processes, the production capacity of which is several times greater than that of a conventional process.

Process Intensification

Equipment		Methods			
With reactions	*Without*	*Multifunctional reactors*	*Hybrid separations*	*Alternative energy sources*	*Other methods*
Rotating disk	Static mixer	Reactive distillation	Membrane absorption	Centrifugal forces	Supercritical fluids
Reactive static mixer	Compact heat exchanger	Reactive extraction	Membrane distillation	Ultrasound	Periodic operation
Catalytic static mixer	Micro-channel heat exchanger	Reactive crystallization	Adsorptive distillation	Solar power	
Monolith	Rotor-stator mixer	Chromatographic reactor		Microwave	
Microreactor	Rotating fixed bed	Periodic separating reactor		Electric fields	
Heat exchanger reactor	Centrifugal adsorber	Membrane reactor		Plasmas	
Supersonic G/L reactor		Reactive extrusion			
Impact jet reactor		Fuel cell			
Rotating fixed bed		Flow reversal reactor			

Figure 5.23. *Classification of equipment and process intensification methods*

Stankiewicz and Moulijn (2000) provided an overview of intensification technologies (equipment and/or methods) (Figure 5.23). The equipment, with or without chemical reactions, includes iconic technologies (rotating disk reactors, static mixers, monolithic reactors, microstructured reactors) and intensive technologies in terms of their transfer characteristics or their ability to combine operations that were previously physically separated. The methods include multifunctional reactors that combine reaction and separation, to push balanced reactions towards productivity and to no longer be limited by thermodynamics. Alternative energy sources (microwave, ultrasound) are considered to be intensive because they focus energy more precisely than mechanical agitation or advection/convection.

However, as exhaustive as such a classification may be, it is not a decision support tool and does not favor any technology for a given problem. A more relevant way of approaching intensification involves analyzing the phenomena that slow down or limit the performance of a piece of equipment or a set of equipment in a process, and implement appropriate strategies. The equipment involves multiple phenomena, combined at different spatial and temporal scales: the slowest processes have the highest characteristic times, which reflects the influence of limiting phenomena. The means of intensification therefore vary according to the nature of the limiting processes.

At the equipment level, 17 potential limitations can be identified (Figure 5.24) and 17 intensification strategies allow for selective action to reduce these limitations, with the relationships of relevance being defined by a connection matrix (Commenge and Falk 2014). Some strategies lead to specific commercially available technologies, others lead to the invention of innovative equipment. The strategies commonly put to use aim to overcome the limitations due to heat and materials transfer (either heat elimination or supply) as well as to shift the thermodynamic limitations (solubility, chemical reaction equilibrium).

At the process level, more complex approaches to synergies, sensitivity analyses, pinch analyses, energy integration, mixed optimization, and process synthesis methods must be used and combined in order to benefit overall from a local performance improvement (Portha *et al.* 2014).

Because of its multidisciplinary nature, process intensification is a method of engineering complexity, which can prove difficult to carry out, particularly as it requires reconsidering the synthesis routes, but which can improve competitiveness.

Despite the confidentiality of many of these developments, in 2019, there were more than 2,000 industrial reactors implementing flux inversion, 300 membrane units for hydrogen separation, 200 reactive distillation units, around one hundred internal wall distillation columns, and several dozen rotating fixed beds and microwave heating units (Harmsen 2019).

The fields of application range from industrial gases to specialty chemicals. More than a simple technological evolution, process intensification induces profound changes in certain fields like pharmaceutical chemistry, where the traditional methods of discontinuous production are called into question, in favor of compact and continuous heat exchanger reactors.

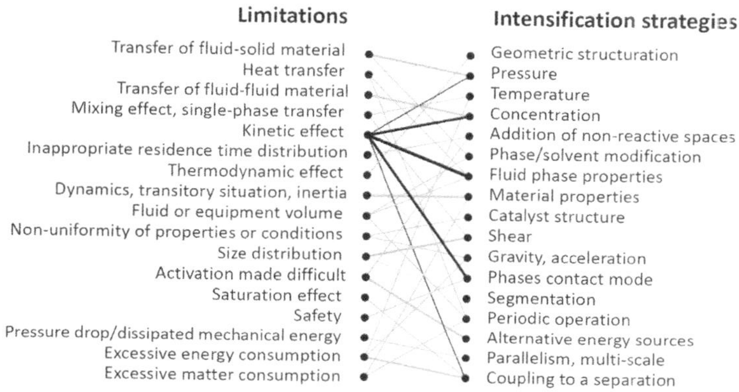

Figure 5.24. *Potential equipment limitations, possible intensification strategies, and relationships of relevance*

For more information

Reay, D., Ramshaw, C., Harvey, A. (2008). *Process Intensification: Engineering for Efficiency, Sustainability and Flexibility*. Butterwoth-Heinemann, Oxford.

Segovia-Hernandez, J.G., Bonnilla-Petriciolet, A. (2016). *Process Intensification in Chemical Engineering: Design Optimization and Control*. Springer International Publishing, New York.

Stankiewicz, A., Van Gerven, T., Stefanidis, G. (2019). *The Fundamentals of Process Intensification*. Wiley-VCH, Weinheim.

Videos

"Basics of Process Intensification – Lecture by Georgios Stefanidis and Andrzej Stankiewicz", 1:10 (accessed July 16, 2019):

https://www.youtube.com/watch?v=QXrQxClahoA

"Continuous Flow Synthesis: Laboratory Approach & Protocol for Scale-up – Lecture by Amol Kulkarni", 33 min. (accessed July 16, 2019):

https://www.labtube.tv/video/continuous-flow-synthesis-laboratory-approach-protocol-for-scale-up

Jean-Marc Commenge, Jean-François Portha, and Laurent Falk

Jean-Marc Commenge

Jean-François Portha

Laurent Falk

Jean-Marc Commenge is a Professor of chemical engineering at ENSIC (French National Chemical Engineering School), linked to the LRGP (Laboratory for Reactions and Chemical Engineering), where he is responsible for the PRIMO research axis (processes, reactors, intensification, membranes, optimization).

jean-marc.commenge@univ-lorraine.fr

Jean-François Portha is an Associate Professor at ENSIC (French National Chemical Engineering School), linked to the LRGP (Laboratory for Reactions and Chemical Engineering), where his work focuses, in particular, on the experimental and numerical characterization of intensified processes (heterogeneous reactors, heat exchangers).

jean-francois.portha@univ-lorraine.fr

Laurent Falk is a Research Director at the CNRS and director of the LRGP (Laboratory for Reactions and Chemical Engineering).

laurent.falk@univ-lorraine.fr

Laboratoire Réaction et Génie des Procédés (UMR CNRS UL 7274). 1, rue Grandville, BP 20451, Nancy Cedex.

Box 5.1. *Process intensification (Jean-Marc Commenge, Jean-François Portha and Laurent Falk)*

Modeling/simulation in chemical engineering

The objectives of modeling are to organize/capitalize on knowledge and to provide "one" representation of reality.

Thus, Minsky (1985) gave the following definition: "An object A is a model of an object B if an observer can use A to answer the questions that interest him about B."

Marquardt (1994) specified that a model is an "abstraction from reality which can be used to represent **certain** aspects of a real process, considered important by the modeler".

These two definitions perfectly underline the fact that we could propose many models from the same process, according to the degree of complexity that we are able to describe or rather, according to the objective that we want to achieve. This invites us to reflect before embarking on what we wish to obtain and the means available for this; only then can we choose the type of model to be used and the associated simulation tool.

Simulation "is a calculation (generally not manual), the resolution of one (or more) coupled equation(s) from input data, possibly under constraints".

It is usual in chemical engineering, as Meyer (2012) pointed out, to differentiate four categories of models, in decreasing order of knowledge and increasing number of experiments to be carried out:

– the pure knowledge model, based solely on theoretical knowledge of the system, which is still difficult to express today and even more difficult to simulate;

– the phenomenological model (also called the "gray box model"), based as much as possible on physical laws, but in which we use empirical correlations for certain aspects, either to simplify the simulation, or for lack of knowledge. We start by writing the material and energy balances as well as the momentum conservation balance (this is the hydrodynamic model), to which we add constitutive equations (thermodynamic models, various correlations) and constraint equations (closing equations, boundary conditions, equilibrium equations, mathematical constraints related to control and optimization, for example);

– the hybrid model, which combines, for example, writing balances and neural networks. It can be used, in particular, for very complex reactive systems such as combustion or biological systems, involving dozens or even hundreds of chemical reactions;

– the empirical model or behavioral model (known as the "black box" model), based solely on experimental information and which can take the form of a transfer function, a neural network, or simply a polynomial.

Meyer classified the models obtained by dimensional analysis in the last category, while stressing that "this approach deserves the full attention of the chemical engineering community".

Various model scales can also be shown in, as in Figure 5.25. So, we find:

– at the level of a manufacturing plant, **modular simulators** such as ProSim, Aspen, ProII, etc. These simulators all include a thermodynamic database (properties of pure substances, models for calculating the properties of mixtures) and a graphical interface enabling the process diagram to be reproduced by arranging the various pieces of equipment that constitute it and connecting them by material flow and sometimes intangible links, particularly in order to optimize the process. Equipment models are generally phenomenological, with a very simplified representation of flows. The results of these simulations consist of characterization of each material flow between the equipment and at the output (essentially temperature, pressure, composition, flow);

– at the level of equipment, we can find the **phenomenological, empirical, or hybrid** models mentioned above;

– between the micronic and millimeter scale, we find the tools of **digital fluid mechanics** or **multiphysics simulation**, particularly CFD and DEM, in which we start by making a mesh of the space to be studied, before solving the Navier-Stokes equation for CFD or representations of the shocks between particles for DEM and the laws of continuity and conservation, even chemical reactions. The results of these simulations take the form of pressure, temperature, velocity field charts, in particular;

– at the molecular level, **molecular modeling** tools, the interest of which in chemical engineering can, for example, be the prediction of properties, transport, in particular (see section 4.2.2.2).

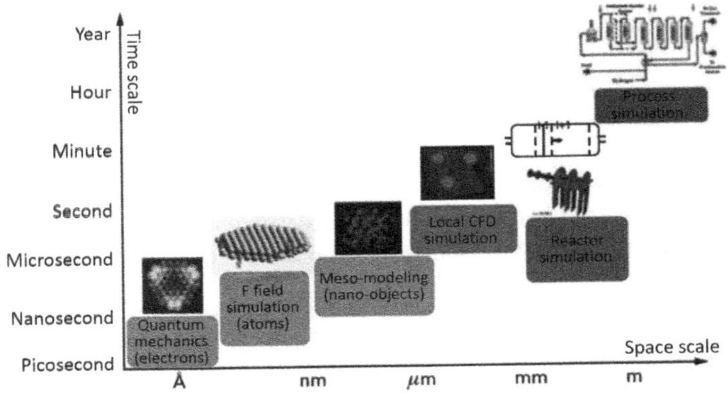

Figure 5.25. *Different modeling scales and simulation in chemical engineering (source: Charpentier (2013)). For a color version of this figure, see www.iste.co.uk/dalpont/process1.zip*

To make these different simulation tools interact with each other (notions of interoperability), the CAPE-OPEN standard has been established. It is managed by COLaN[1].

For more information

Joulia, X. (2008). *Simulateurs de procédés*. Techniques de l'Ingénieur, Saint-Denis.

Meyer, X.M. (2012). *Modélisation en génie des procédés*. Techniques de l'Ingénieur, Saint-Denis.

Marie Debacq

Marie Debacq was a lecturer at Cnam, where her research work concerned the experimental study and modeling of multiphase reactors, in particular rotary kilns. Recently, she has been in charge of the technology platform at AgroParisTech.

Box 5.2. *Modeling/simulation in chemical engineering (Marie Debacq)*

Pilot projects to study unit operations in chemical engineering

Whether for educational purposes (to study an operation or learn to operate equipment) or R&D (development of a completely new process, transposal of a preliminary process in the laboratory to an industrial scale, detailed study of an operation), academic and industrial laboratories use pilot facilities that are smaller than industrial equipment yet large enough for hydrodynamic and thermal aspects, as well as transfer issues, to be present and for their effect on the efficiency of the process to be studied.

The following photos and videos show some examples of such pilot facilities.

1 http://www.colan.org.

Figure 5.26. *Multipurpose reactor used for teaching and research (photograph by Clément Haustant, 2019). For a color version of this figure, see www.iste.co.uk/dalpont/process1.zip*

Figure 5.27. *Pilot rotary kiln designed for research work (photograph by Phahath Thammavong, 2009). For a color version of this figure, see www.iste.co.uk/dalpont/process1.zip*

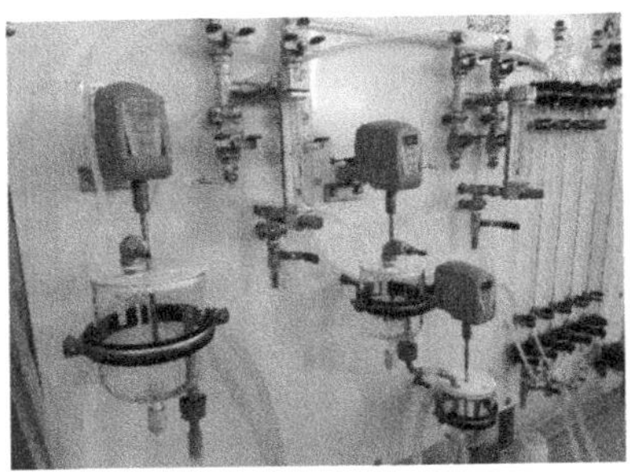

Figure 5.28. *Practical work bench for "reactors": Comparison of conversion rates at the outlet of different reactors and study of residence time distribution (photograph by Clément Haustant, 2019). For a color version of this figure, see www.iste.co.uk/dalpont/process1.zip*

Figure 5.29. *Practical work bench for "pressure drop" (photograph by Clément Haustant, 2019). For a color version of this figure, see www.iste.co.uk/dalpont/process1.zip*

Figure 5.30. *Filter press pilot plant (photograph by Clément Haustant, 2019). For a color version of this figure, see www.iste.co.uk/dalpont/process1.zip*

Videos (courtesy of Pignat)

Continuous crystallization:

https://youtu.be/27Gc5b549Rc

Automatically controlled continuous distillation:

https://youtu.be/koob0mGjgkE

Multitubular heat exchangers:

https://youtu.be/rQ8rJj_Zf3s

Single effect evaporator:

https://youtu.be/cu3VYKCfQ_M

Marie Debacq

Marie Debacq was a lecturer at Cnam, where her research work concerned the experimental study and modeling of multiphase reactors, in particular rotary kilns. Recently, she has been in charge of the technology platform at AgroParisTech.

Box 5.3. *Pilot projects to study unit operations in chemical engineering (Marie Debacq)*

Where does the Logarithmic Mean Temperature Difference (LMTD) come from?

We are going to consider the simple case of a coaxial heat exchanger, mentioned in section 5.4. We will carry out an enthalpy balance calculation on a section of the heat exchanger $\mathrm{d}x$ to understand where the formula $\phi = HS\Delta T_{ml}$, given in this, comes from.

Let us consider the assumptions:

– the exchanger is thermally insulated;

– there is negligible conduction along the axis;

– the specific heats of hot and cold fluids remain constant;

– H is constant over the entire length of the exchanger.

The transverse heat flow on this section, corresponding to the heat transferred from the hot fluid to the cold fluid through the wall of the tube, is $\mathrm{d}\phi = H\mathrm{d}S(T_c - T_f)$. S is the heat transfer surface between the two fluids separated by the wall of the tube. H is the heat transfer coefficient and its unit is kW m^{-2} K^{-1}. In the absence of heat loss, the heat flow $\mathrm{d}\phi$ corresponds to the heat given off by the hot fluid $-Q_{mc}.c_{pc}\mathrm{d}T_c$, as well as to the heat gained by the cold fluid $Q_{mf}.c_{pf}\mathrm{d}T_f$, over the length $\mathrm{d}x$ of the section. In other words, since there is no heat loss, all the heat given off by the hot fluid is transferred to the cold fluid.

c_{pc} and c_{pf} denote the mass heat of the hot and cold fluids, respectively, and are expressed in J mol^{-1} K^{-1}. Q_{mc} and Q_{mf} denote the mass flow rates of each of the two fluids in kg s^{-1}.

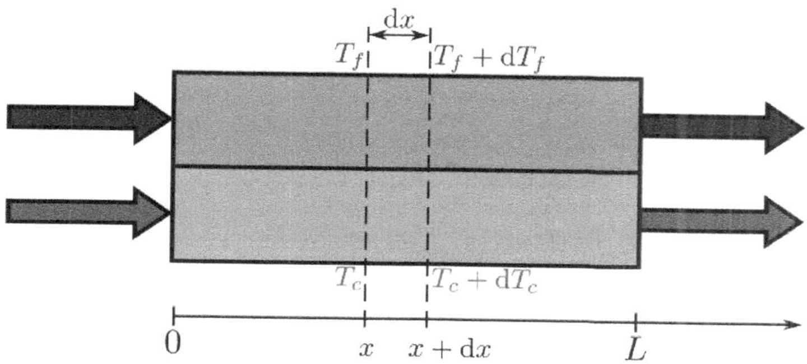

Figure 5.31. *Illustration of the energy balance on an elementary section of the tube of a co-current heat exchanger. For a color version of this figure, see www.iste.co.uk/dalpont/process1.zip*

We thus have:

$$d\phi = HdS(T_c - T_f) = -Q_{mc}.c_{pc}dT_c = Q_{mf}.c_{pf}dT_f$$

where:

$$dT_c = -\frac{d\phi}{Q_{mc}.c_{pc}} \quad et \ dT_f = -\frac{d\phi}{Q_{mf}.c_{pf}}$$

$$dT_c - dT_f = d(T_c - T_f) = -\left(\frac{1}{Q_{mc}.c_{pc}} + \frac{1}{Q_{mf}.c_{pf}}\right)d\phi$$

As $d\phi = HdS(T_c - T_f)$, we have:

$$d(T_c - T_f) = -\left(\frac{1}{Q_{mc}.c_{pc}} + \frac{1}{Q_{mf}.c_{pf}}\right)HdS(T_c - T_f)$$

where:

$$\frac{d(T_c - T_f)}{T_c - T_f} = -\left(\frac{1}{Q_{mc}.c_{pc}} + \frac{1}{Q_{mf}.c_{pf}}\right)HdS$$

If we integrate this expression from $S = 0$ to S, that is to say between the input and the output of the exchanger, we obtain:

$$\ln\left[\frac{(T_c - T_f)_s}{(T_c - T_f)_e}\right] = -\left(\frac{1}{Q_{mc}.c_{pc}} + \frac{1}{Q_{mf}.c_{pf}}\right) HS$$

However, the total exchanged flow ϕ can be expressed according to the input and output temperatures of the fluids in the exchanger:

$$\phi = Q_{mc}.c_{pc}(T_{ce} - T_{cs}) = Q_{mf}.c_{pf}(T_{fs} - T_{fe})$$

where:

$$\ln\left(\frac{T_{cs} - T_{fs}}{T_{ce} - T_{fe}}\right) = -\left(\frac{T_{ce} - T_{cs}}{\phi} + \frac{T_{fs} - T_{fe}}{\phi}\right) HS = \left[(T_{cs} - T_{fs}) - (T_{ce} - T_{fe})\right]\frac{HS}{\phi}$$

For a coaxial co-current tube exchanger, the total exchanged flow can therefore be expressed according to the input and output temperatures of the two fluids, the transfer coefficient, and the transfer surface:

$$\phi = HS\frac{\left[(T_{cs} - T_{fs}) - (T_{ce} - T_{fe})\right]}{\ln\left(\dfrac{T_{cs} - T_{fs}}{T_{ce} - T_{fe}}\right)}$$

In the case of a counter-current coaxial heat exchanger, the same calculations can be made from the flow expression, which is expressed, in this case, according to:

$$\mathrm{d}\phi = H\mathrm{d}S(T_c - T_f) = -Q_{mc}.c_{pc}\mathrm{d}T_c = -Q_{mf}.c_{pf}\mathrm{d}T_f$$

Indeed, for a counter-current exchanger, the temperature variation of the cold fluid $\mathrm{d}T_f$ becomes negative when the transfer surface is increased by the quantity $\mathrm{d}S$. This leads to the following expression of the total exchanged flow:

$$\phi = HS\frac{\left[(T_{ce} - T_{fs}) - (T_{cs} - T_{fe})\right]}{\ln\left(\dfrac{T_{ce} - T_{fs}}{T_{cs} - T_{fe}}\right)}$$

For these two types of tubular heat exchangers, ΔT_1 can denote the temperature difference between the two fluids in $x = 0$ and ΔT_2 can denote this difference in $x = L$, with L being the total length of the exchanger. ϕ is then expressed by the general formula:

$$\phi = HS\frac{\Delta T_2 - \Delta T_1}{\ln\left(\dfrac{\Delta T_2}{\Delta T_1}\right)} = HS\frac{\Delta T_1 - \Delta T_2}{\ln\left(\dfrac{\Delta T_1}{\Delta T_2}\right)} = HS\Delta T_{ml}$$

with ΔT_{ml} the logarithmic mean temperature difference of the two fluids, introduced in section 5.4.2, and reflecting the fact that the temperature difference between the two fluids is not constant throughout the exchanger.

DIGITAL APPLICATION.– In a plant, at the bottom of a distillation column, we want to be able to cool products using a counter-current tubular heat exchanger that runs on crude oil. A flow rate of 29.6 kg s^{-1} of product must be cooled from 420 to 380 K; crude oil is available at 295 K and its temperature must not exceed 330 K. The specific heats for the product and the oil are respectively 2,200 J kg^{-1} K^{-1} and 1,986 J kg^{-1} K^{-1}. We know that the system is characterized by an overall transfer coefficient H equal to 433.8 J s^{-1} m^{-2} K^{-1}. We can recall, as seen in section 5.4, that the overall transfer coefficient takes into account the convection on either side of the exchange wall (in the hot fluid and in the cold fluid) and the conduction in the material constituting the wall. To calculate this kind of coefficient, we must therefore take into account, in particular, the thermophysical properties of fluids such as thermal conductivity, heat capacity, viscosity, etc. The thermal conductivity of the material of the wall also plays a role. Using the data available, we find the flow rate of crude oil involved in this exchanger, as well as the transfer surface that characterizes it.

From the equations seen above, we can write:

$$\phi = HS\Delta T_{ml} = -Q_{mc}.c_{pc}\Delta T_c = -Q_{mf}.c_{pf}\Delta T_f$$

hence the crude oil flow rate:

$$Q_{mf} = \frac{Q_{mc}.c_{pc}\Delta T_c}{c_{pf}\Delta T_f} = \frac{29,6 \times 2200 \times (420 - 380)}{1986 \times (330 - 295)} = 37,47 \text{ kg s}^{-1}$$

and the transfer surface of the exchanger:

$$S = \frac{\phi}{H\Delta T_{ml}} = \frac{-Q_{mc}.c_{pc}\Delta T_c}{H\Delta T_{ml}} = \frac{29,6 \times 2200 \times (420 - 380)}{433,8 \times \dfrac{(420-330)-(380-295)}{\ln\left(\dfrac{420-330}{380-295}\right)}} = 68,6 \text{ m}^2$$

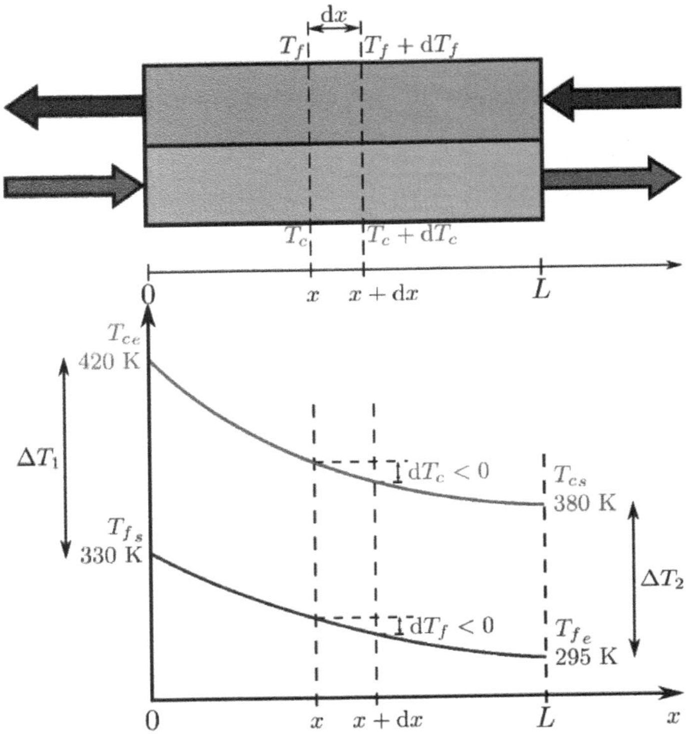

Figure 5.32. *Illustration of energy balance on an elementary section of the tube of a counter-current exchanger and temperature profiles in the exchanger. For a color version of this figure, see www.iste.co.uk/dalpont/process1.zip*

The relationship $\phi = HS\Delta T_{ml}$ tells us that the greater the temperature difference between the hot fluid and the cold fluid, the greater the heat flow transferred; in other words, the more efficient the transfer. This seems logical because the temperature difference is the driving force of the transfer. Note that in a simple tubular exchanger, the transferred heat flow is always higher in the case of counter-current operations than in the case of co-current operations because the ΔT_{ml} is higher in the latter. For example, in the case of our exchanger:

– counter-current: $\Delta T_{ml} = \dfrac{(420-330)-(380-295)}{\ln\left(\dfrac{420-330}{380-295}\right)}$ 87.48 K;

– co-current: $\Delta T_{ml} = \dfrac{(420-295)-(380-330)}{\ln\left(\dfrac{420-295}{380-330}\right)} = 81.85$ K.

Céline Houriez

Céline Houriez is a research fellow at the *École des mines de Paris*. She is attached to the *Centre Thermodynamique des Procédés*, (CTP) where her work is mainly dedicated to predicting thermophysical properties by molecular simulation.

Box 5.4. *Where does the Logarithmic Mean Temperature Difference (LMTD) come from? (Céline Houriez)*

5.8. Glossary

Adiabatic: takes place without heat exchange to and from an external source.

Computational Fluid Dynamics (CFD): digital fluid mechanics, simulation tool based on solving the Navier-Stokes equations and turbulence models.

Continuous: a type of operation where the equipment is permanently supplied by one or more incoming flows and also with continuous evacuation of the outgoing flow or flows.

Counter-current: a principle of contact between two flows flowing in opposite directions, making it possible to maintain a difference in temperature and/or composition favoring transfers.

Discontinuous: a type of operation where the equipment is, in a first phase, supplied with material, then undergoes various operating phases (heating, separation, cooling, reaction, etc.) before discharging the residual material that it contains.

Discrete Element Method (DEM): numerical method for simulating the movement of a large number of small particles and possibly the heat transfers between these particles.

Height equivalent to a theoretical plate (HETP): packing height corresponding to a theoretical stage in a separation column (distillation, absorption, liquid-liquid extraction).

Number of equilibrium stages (NES): number of thermodynamic equilibrium stages allowing for a given separation to be carried out.

Residence time distribution (RTD): method for diagnosing faults and modeling macroscopic flows in equipment.

5.9. References

Anglaret, P. (1998). Technologie génie chimique. Centre régional de documentation pédagogique de l'académie d'Amiens.

Charpentier, J.C. (2013). *Génie des procédés, développement durable et innovation. Enjeux et perspectives*. Techniques de l'Ingénieur, Saint-Denis.

Commenge, J.-M., Falk, L. (2014). Methodological framework for choice of intensified equipment and development of innovative technologies. *Chemical Engineering and Processing: Process Intensification*, 84, 109–127.

Coulson, J.M., Richardson, J.F. (2002). *Coulson and Richardson's Chemical Engineering*, 5th edition. Butterworth Heinemann, Oxford.

Harmsen, J. (2019). *Industrial Process Scale-up: A Practical Innovation Guide from Idea to Commercial Implementation*, 2nd edition. Elsevier, Amsterdam.

Humphrey, J.L. (2001). *Procédés de séparation. Techniques, sélection, dimensionnement.* Dunod, Paris.

Jachuck, R.J.J., Ramshaw, C. (1994). Process intensification: Heat transfer characteristics of tailored rotating surfaces. *Heat Recovery Systems and CHP*, 14(5), 475–491.

Levenspiel, O. (1999). *Chemical Reaction Engineering*. Wiley, New York.

McCabe, W.L., Smith, J.C. (2001). *Unit Operations of Chemical Engineering*, 6th edition. McGraw-Hill, New York.

Midoux, N. (1985). *Mécanique et rhéologie des fluides en génie chimique.* Lavoisier, Paris.

Midoux, N. (1996). Agitation – Mélange monophasique. Polycopié Ensic, Nancy.

Perry, R.H., Green, D.W. (1997). *Perry's Chemical Engineers' Handbook*, 7th edition. McGraw-Hill, New York.

Portha, J.-F., Falk, L., Commenge, J.-M. (2014). Local and global process intensification. *Chemical Engineering and Processing: Process Intensification*, 84, 1–13.

Ramshaw, C. (1993). The opportunities exploiting centrifugal fields. *Heat Recovery Systems and CHP*, 13(6), 493–513.

Roustan, M. (2005). *Agitation. Mélange – Caractéristiques des mobiles d'agitation*. Techniques de l'Ingénieur, Saint-Denis.

Stankiewicz, A., Moulin, J.A. (2000). Process intensification: Transforming chemical engineering. *Chemical Engineering Progress*, 22–34.

Villermaux, J. (1993). *Génie de la réaction chimique*. Lavoisier, Paris.

List of Authors

Laurent BASEILHAC
Arkema
Paris
France

Jean-Pierre DAL PONT
SECF
Paris
France

Marie DEBACQ
Cnam – AgroParisTech
Paris
France

Alain GAUNAND
École des mines de Paris
France

Céline HOURIEZ
École des mines de Paris
France

Vincent LAFLÈCHE
École des mines de Paris
France

Willi MEIER
DECHEMA
Frankfurt
Germany

Michel ROYER
Consultant
Chaingy
France

June C. WISPELWEY
AIChE
New York
USA

Index

A

ADEME, 64
alginates, 95
analysis
  Pareto, 10
artificial intelligence, 66
aspirin, 9, 89, 97–99, 102, 110–113
assessment
  lifecycle (LCA), 33, 49

B

barcode, 60
biofuels, 68
blockchain, 37, 62, 63
business plan, 6, 10

C

catalysis, catalysts, 72, 140
chemical kinetics, 129, 136, 149, 198, 206, 207
circular economy, 65
cobotics, 66
coffee, 89, 99–102
communicating objects, 42
communicative street furniture, 44
company, 1
connected city, 42
COP21, 25, 68
cryptocurrencies, 62
cycle, 7, 23, 33, 49, 50, 52, 53, 74, 116, 123, 124
  water, 23

D, E

distillation, 79, 108, 157, 159–162, 164–169, 182, 190, 210, 221, 224
Earth, 17
ecotoxicology, 53
Edison, Thomas, 2, 4
endocrine disruptors, 54
energy conservation, 119, 146, 164, 192, 218
engineering, 15, 83, 108, 109, 111, 136, 169, 208, 212, 215, 224
  chemical, 109, 111
  product, 76, 113
enterprise (*see also* company),
  flows, 13
  industrial, 4
environmental impacts, 49

F

factories, 1
Fayol, Henri, 3

Federec, 64, 67
flexibility, 41
food chain, 56, 57
Ford, Henry, 3
formulation, 9, 35, 36, 60, 76, 77, 79, 83, 98, 99, 101, 110, 113, 122, 123
functionality, 78

G, H

gelatin, 92–94
governance, 14
hazard, 54
household waste, 63, 64
hydro-isomerization, 69
hydrocolloids, 95
hydrotreatment, 69

I, L, M

Industrial Revolution, 1, 2
industry strategy, 6
Internet of Things (IoT), 44
lifecycle, 56
  product, 7
manufacturing industries, 2
material balance, 164, 198
microencapsulation, 94, 95, 104

P, Q

plastics, 65
process modeling, 51
product stewardship, 58
products
  carbon dioxide, 25, 104, 126
  chemical, 21, 53–55, 57, 85, 88, 89, 104
  composite, 77
  researchers in the chemical industry, 104
  specialty, 77, 81
QR code, 42, 60

R

recycling entrepreneurs, 63
Reengineering the Corporation, 11
reverse engineering, 82
Reynolds number, 133, 171
RFID, 60, 61
risk, 54
robotics, 66

S

sensors, 42
Smart City, 21, 39, 40, 42, 45–49
Smith, Adam, 2
starch, 90–92, 98
systemic vision, 11

T

Taylor, Frederick Winslow, 3
thermodynamics, 114
toxicology, 53
traceability, 59
transfer, 127, 131, 136
  heat, 127, 130
  mass, 128, 130
  momentum, 128, 131
transition, 23, 65, 66, 68, 76, 119
  digital, 66
  energy, 65
trickle-bed, 73

U, W

useful property, 79
water-energy-food-climate nexus, 22, 24
wrapping, 84

Summary of Volume 2

Foreword 1
Laurent BASEILHAC

Foreword 2
Vincent LAFLÈCHE

Foreword 3
June C. WISPELWEY

Introduction
Jean-Pierre DAL PONT and Marie DEBACQ

Chapter 1. Bio-industry in the Age of the Transition to Digital Technology: Significance and Recent Advances
Philippe JACQUES

- 1.1. Introduction
- 1.2. Diversity of products and applications
  - 1.2.1. Fermentations in agri-food
  - 1.2.2. Biomass-based products
  - 1.2.3. Metabolite-based products
- 1.3. Traditional process for developing a product of industrial microbiology
- 1.4. Strain selection and optimization
  - 1.4.1. Evolution of strain screening techniques
  - 1.4.2. Evolution of genetic modification technologies, from random mutagenesis to CRISPR-Cas9 technology
- 1.5. Production and purification processes
  - 1.5.1. Needs of microorganisms
  - 1.5.2. Production processes

1.5.3. Downstream fermentation processes
(downstream processing, DSP)
1.5.4. Coupled procedures
1.5.5. Microfluidic intake (scale-up/scale-down)
1.5.6. Process intensification
1.6. Innovative concepts
1.6.1. Biofilm reactors
1.6.2. Mixed cultures and cascades of microorganisms
1.7. Towards a digital bio-industry
1.8. Acknowledgements
1.9. Glossary
1.10. References

Chapter 2. Hydrogen Production by Steam Reforming
Marie BASIN, Diana TUDORACHE, Matthieu FLIN, Raphaël FAURE and Philippe ARPENTINIER

2.1. The industrial production of hydrogen
2.1.1. The processes of hydrogen production
2.1.2. Natural gas steam reforming
2.2. Problems and operational constraints in steam reforming units
2.2.1. Tube temperature and lifetime
2.2.2. Catalyst deactivation
2.2.3. Corrosion by metal dusting
2.2.4. Flexibility in raw materials of steam reforming units
2.3. Recent industrial developments responding to global warming
2.3.1. The role of hydrogen in energy transition
(Hydrogen Council 2017)
2.3.2. CO_2 capture in hydrogen production units
2.3.3. The exchanger reactor ("zero steam")
2.3.4. Current research interests
2.3.5. Other means to provide reaction heat
(currently in development)
2.4. References

Chapter 3. Industrialization: From Research to Final Product
Jean-Pierre DAL PONT

3.1. Anatomy of a process
3.2. Process evaluation
3.3. Industrialization process
3.3.1. The foundations of industrialization
3.3.2. Realization (project) engineering

3.4. The concept of the industrial project
3.5. Typical organization of an industrial project
3.6. The stages of an industrial project from the engineering perspective – validations
3.7. The tools of engineering project management – related activities
3.7.1. Process conceptualization: making it visible
3.7.2. Project management
3.7.3. Reporting – executive summary
3.7.4. Other concepts
3.8. Process intensification (PI) – miniaturization
3.9. Investment/sales coupling – modular construction
3.10. Circular industrial economy – platforms – centralization – decentralization
3.11. Overseas operations – technology transfer
3.12. Conclusion
3.13. Boxes
3.14. References

Chapter 4. Operations

Jean-Pierre DAL PONT

4.1. The industrial tool seen by flows and Enterprise Resource Planning (ERP)
4.2. The supply chain
4.3. The typology of the means of production: VAT analysis
4.4. The anatomy of a plant
4.5. Operations management systems, the push for excellence
4.5.1. A brief history of industrial operations management
4.5.2. Toyotism
4.6. Costing-based profitability analysis (CO-PA): measure of performance and steering tool
4.6.1. Product cost
4.6.2. Margins
4.6.3. The breakeven point: the absorption of fixed costs
4.6.4. The infernal spiral of fixed costs
4.6.5. Observations on margins
4.7. The plant: performance measurement and score cards
4.8. Change management
4.8.1. Processes: system integrity and robustness
4.8.2. Human aspects and climate of trust
4.8.3. Knowledge management and core competencies
4.8.4. Continuous improvement and the search for innovation

4.8.5. The search for technological breakthrough and innovation
4.8.6. Operations abroad
4.8.7. What about tomorrow?
4.9. References

Chapter 5. The Enterprise and the Plant of the Future at the Age of the Transition to Digital Technology
Jean-Pierre DAL PONT

5.1. From one Industrial Revolution to the next Industrial Revolution
5.1.1. The First Industrial Revolution (1712–1860): steam, a source of energy
5.1.2. The Second Industrial Revolution (1860–1960): from crafts to industrial enterprise
5.1.3. The Third Industrial Revolution (1960–1990): the rise of industrial computing
5.1.4. The Fourth Industrial Revolution (1990–present)
5.2. Artificial intelligence (AI): deep learning and machine learning
5.3. Big Data
5.3.1. Characterization
5.4. Digital tools and technologies for industrial enterprise
5.4.1. Products, innovation, management
5.4.2. New tools
5.4.3. Digital twins
5.4.4. Engineering revisited
5.4.5. 3D (three-dimensional) printer or additive manufacturing
5.4.6. Robots, robotics, exoskeletons
5.4.7. Drones
5.4.8. Operations management
5.5. Boxes
5.6. References

Chapter 6. And Tomorrow...
Jean-Pierre DAL PONT

6.1. The beginning of an epic: business, science, technology, the leap forward
6.2. Artificial intelligence (AI) and economic channels
6.2.1. Medicine and health
6.2.2. The water-energy-food-climate nexus
6.2.3. Intelligent electrical network (Smart Grid)
6.2.4. Artificial Intelligence and Smart City

6.3. Artificial intelligence and the consumer
6.4. Artificial intelligence, environment and human factor
6.5. The human at the heart of the device, at the heart of the system
6.5.1. Humans and robots
6.6. System robustness, resilience and fragility
6.7. GAFA: concerns, fears, myths and phantasms
6.8. Industrial companies in the face of digital technology
6.8.1. Cybercrime and uberization
6.8.2. Software hybridization
6.8.3. After Fordism and Toyotism, Teslism?
6.8.4. Business and governance: products
6.8.5. The chemical engineer, the project management
6.9. Towards a Black Box Society?
6.10. Conclusion
6.11. Box
6.12. References

Other titles from

iSTE

in

Chemical Engineering

2020

BELAADI Salah
Thermodynamic Processes 1: Systems without Physical State Change
Thermodynamic Processes 2: State and Energy Change Systems

SCHAER Éric, ANDRÉ Jean-Claude
Process Engineering Renewal 1: Background and Training
Process Engineering Renewal 2: Research

2019

HARMAND Jérôme, LOBRY Claude, RAPAPORT Alain, SARI Tewfik
Chemostat and Bioprocesses Set
Volume 3 – Optimal Control in Bioprocesses: Pontryagin's Maximum Principle in Practice

2018

LOBRY Claude
Chemostat and Bioprocesses Set
Volume 2 – The Consumer–Resource Relationship: Mathematical Modeling

2017

HARMAND Jérôme, LOBRY Claude, RAPAPORT Alain, SARI Tewfik
Chemostat and Bioprocesses Set
Volume 1 – The Chemostat: Mathematical Theory of Microorganims Cultures

2016

SOUSTELLE Michel
Chemical Thermodynamics Set
Volume 5 – Phase Transformations
Volume 6 – Ionic and Electrochemical Equilibria
Volume 7 – Thermodynamics of Surfaces and Capillary Systems

2015

SOUSTELLE Michel
Chemical Thermodynamics Set
Volume 1 – Phase Modeling Tools
Volume 2 – Modeling of Liquid Phases
Volume 3 – Thermodynamic Modeling of Solid Phases
Volume 4 – Chemical Equilibria

2014

DAL PONT Jean-Pierre, AZZARO-PANTEL Catherine
New Approaches to the Process Industries: The Manufacturing Plant of the Future

HAMAIDE Thierry, DETERRE Rémi, FELLER Jean-François
Environmental Impact of Polymers

2012

DAL PONT Jean-Pierre
Process Engineering and Industrial Management

2011

SOUSTELLE Michel
An Introduction to Chemical Kinetics

2010

SOUSTELLE Michel
Handbook of Heterogenous Kinetics

Printed and bound by CPI Group (UK) Ltd, Croydon, CR0 4YY
16/09/2026
14963077-0003